Technology, Urban Space and the Networked Community

Saswat Samay Das • Ananya Roy Pratihar
Editors

Technology, Urban Space and the Networked Community

palgrave
macmillan

Editors
Saswat Samay Das
Indian Institute of Technology Kharagpur
Kharagpur, India

Ananya Roy Pratihar
Insititute of Management and
Information Science
Bhubaneswar, India

ISBN 978-3-030-88808-4 ISBN 978-3-030-88809-1 (eBook)
https://doi.org/10.1007/978-3-030-88809-1

This Palgrave Macmillan imprint is published by the registered company Springer Nature Switzerland AG.
The registered company address is: Gewerbestrasse 11, 6330 Cham, Switzerland

Contents

Notes on Contributors

Ian Buchanan is an Australian cultural theorist, serving as director of the Institute for Social Transformation Research based at University of Wollongong, Australia. He is the author of the *Oxford Dictionary of Critical Theory* and the founding editor of the international journal *Deleuze Studies*. He is also the editor of four book series: *Deleuze Connections*, *Critical Connections*, *Plateaus* and *Deleuze Encounters*.

Rick Dolphijn is an associate professor of Media and Culture Studies at Utrecht University, Netherlands. He has recently written *Michel Serres and the Crises of the Contemporary* (2019), authored *Foodscapes: Towards a Deleuzian Ethics of Consumption* (2004) and *New Materialism: Interviews & Cartographies* (with Iris van der Tuin, 2012), and edited (with Rosi Braidotti) *This Deleuzian Century: Art, Activism, Life* (2014).

Aragorn Eloff is an independent researcher and a co-founder of the Institute for Critical Animal Studies in Africa (www.criticalanimalstudies.org/icas-africa/). He is the co-editor of *Deleuze and Anarchism*.

Samir Gandesha is an associate professor in the Department of the Humanities and director of the Institute for the Humanities at Simon Fraser University, Canada. He has recently edited *Spectres of Fascism:*

Historical, Theoretical and International Perspectives (2020), and also co-edited *Aesthetic Marx* (2018) and *Arendt and Adorno: Political and Philosophical Investigations* (2012).

Gary Genosko is Professor of Sociology at Lakehead University, Canada. He is the editor of *The Guattari Reader* and *Deleuze and Guattari: Critical Assessments of Leading Philosophers*, and author of *Baudrillard and Signs*, *Undisciplined Theory* (1998) and *McLuhan and Baudrillard: The Masters of Implosion* (1999).

Emine Görgül is an associate professor and former vice-chair at Istanbul Technical University, School of Architecture, Turkey. She has authored numerous articles and book chapters both in Turkish and in English, on design theory and criticism as well as design education and innovative interventions.

Chantelle Gray is an Associate Professor in the School of Philosophy at North-West University, South Africa. Her books include *Deleuze and Anarchism* (2019), co-edited with Aragorn Eloff, and *Anarchism After Deleuze and Guattari: Fabulating Futures* (2022).

Vinzenz Hediger is Professor of Cinema Studies at Goethe Universität Frankfurt, Germany, and he has co-edited *The State of Post-Cinema: Tracing the Moving Image in the Age of Digital Dissemination* (Palgrave Macmillan, 2016) and *Films That Work: Industrial Film and the Productivity of Media: Studies in the Visual Culture of the Industrial Film* (2014).

Tony See Sin Heng is the editor of *Deleuze and Buddhism* (Palgrave Macmillan, 2016). He is Associate Lecturer in Political Science and International Relations at the University of London (UOL) Programme in the Singapore Institute of Management-Global Education (SIMGE) Centre.

Pramod K. Nayar teaches at the University of Hyderabad, India, and is the author of *Human Rights and Literature* (Palgrave Macmillan US, 2016), *The Indian Graphic Novel* (2016), *The Extreme in Contemporary Culture* and *Bhopal's Ecological Gothic*. Forthcoming works include *Ecoprecarity* (Routledge) and *Brand Postcolonial* (de Gruyter).

S. L. Revoy is a doctoral candidate in the Cultural Studies Programme at Queen's University, Kingston, Canada. His current research interrogates the philosophy, politics and ethics of software design and cybernetic theory, especially focusing on the ways in which these domains affect the development of surveillant technologies and the production of subjectivity.

Janae Sholtz is associate professor of Philosophy and a coordinator of Women's and Gender Studies at Alvernia University, USA. She is the author of *The Invention of a People: Heidegger and Deleuze on Art and the Political* (2015) and co-editor of *Deleuze and the Schizoanalysis of Feminism* (2019).

List of Figures

1

Biopolitics, Discipline and Governmentality

Ian Buchanan

But this is, after all, the circle of struggle and truth, that is to say, precisely, of philosophical practice.
—Michel Foucault, *Security, Territory, Population: Lectures at the Collège de France 1977–1978* (3)

In 2020 as the global COVID-19 pandemic unfolded, no concept in the cultural theory arsenal was more prominent or more widely deployed in the various attempts that were made to write about the many issues the pandemic raised than Michel Foucault's concept of biopolitics. In large part, though, this owed more to later interpretations of Foucault, particularly Agamben and to a lesser extent Esposito, than it did to his own work. In many ways the various governmental responses (locally, nationally, and internationally) to the pandemic were a perfect illustration of what Foucault wanted to draw our attention to when he first conceived

I. Buchanan (✉)
Institute for Social Transformation Research, University of Wollongong, Wollongong, NSW, Australia
e-mail: ibuchana@uow.edu.au

S. S. Das, A. R. Pratihar (eds.), *Technology, Urban Space and the Networked Community*, https://doi.org/10.1007/978-3-030-88809-1_1

of the concept of biopolitics, but very often too the concept was used by commentators to name a situation that would have been more accurately accounted for by Foucault's earlier work on discipline.

Biopolitics is often treated as a continuation of discipline by other means, but Foucault himself did not see it this way. Biopolitics represents something new in the way power is both defined and administered. Biopolitics does not necessarily supersede discipline, or render it obsolete, but it does add a new layer or strata to the operation of power that demands separate analysis because it brings to bear new functions and yields new types of knowledge. However, in spite of its obvious relevance and applicability to the analysis of the global COVID-19 pandemic, I do not believe the prominence of Foucault's concept in 2020 was due solely to the exigencies of the pandemic or the apparent ready-to-handness of the concept itself. I think rather it reflected the growing recognition in cultural theory today that biopolitics is *the* concept for our time. More than any other concept it provides a model to explain the operations of the largely invisible but nevertheless potent essential working parts of power in the present historical era, which Deleuze with characteristic boldness called 'control society' (a phrase he poached from William Burroughs). An abundance of new critical work on such diverse topics as algorithms, asylum seekers, eating disorders, sexuality, and so on points to this conclusion. As Deleuze once said of difference, today biopolitics is in the air.

Foucault's method starts from the presupposition that there are no universals—as Deleuze and Guattari put it, universals explain nothing; they must be explained. "And then I put the question to history and historians: How can you write history if you do not accept a priori the existence of things like the state, society, the sovereign, and subjects?"[1] Foucault's goal here isn't to use history to challenge the universality of universals, for instance by showing that in fact not all prisons are the same, or that madness is differently conceived throughout history, and so on. Nor is he interested in presupposing universals like the idea of the prison or the concept of madness and then showing how real-world examples have deviated from these models. He wants to focus on specific

[1] Foucault (2008: 3).

practices and use them to build his histories of such things as prisons and asylums. "In other words, instead of deducing concrete phenomena from universals, or instead of starting with universals as an obligatory grid of intelligibility for certain concrete practices, I would like to start with these concrete practices and, as it were, pass these universals through the grid of these practices."[2] In starting with techniques the aim isn't to discover something that was hidden, since in fact they were never hidden, nor to ridicule and reject something that now seems not merely outmoded but thoroughly wrongheaded, such as trepanation as a treatment for mental disorders.

Foucault's aim, rather, is to grasp how a dispersed set of conjunctions are able to hold together "a whole set of practices—from the moment they become coordinated with a *regime of truth*—[and thereby] make what does not exist (madness, disease, delinquency, sexuality, etc.), nonetheless become something, something that however continues to not to exist".[3] The object of Foucault's inquiries, then, is the regime of truth, not the array of techniques he dredges out of the murky depths of the archives; the techniques he brings to light are simply the means of posing the question: what holds all these techniques together? His assumption is not that they are held together, I hasten to add, but that it might be possible to see that somehow they are interrelated. The regime of truth can be understood as the set of conditions that must obtain for it to be possible to pronounce statements that are taken to be true. This is not a matter of determining which statements are true and which statements are false, it runs deeper than that because it refers to a way of thinking, as opposed to a discrete set of thoughts. Foucault's prime examples are drawn from the histories of madness, medicine, prisons, and government.

Foucault introduced the concept of biopolitics in his March 17, 1976, lecture at the Collège de France, which concluded the lecture series for the academic year 1975–1976 posthumously published under the title *Society Must be Defended*.[4] Foucault begins his lecture by noting that "one

---

[2] Foucault (2008: 3).

[3] Foucault (2008: 19 emphasis added).

[4] This lecture is effectively a first draft of the final chapter of the first volume of his history of sexuality project, which was published in 1976 under the title *La Volonté de savoir* (*The History of Sexuality: An Introduction*). It is also worth noting that his book on 'disciplinary society' (as Foucault calls it)

of the basic phenomena of the nineteenth century was what might be called power's hold over life". He goes onto to clarify that what he means by this is "the acquisition of power over man insofar as man is a living being, that the biological came under State control, that there was at least a certain tendency that leads to what might be termed State control of the biological".[5] In line with this tendency, he says, "we see something new emerging in the second half of the eighteenth century: a new technology of power, but this time *it is not disciplinary*. This technology of power does not exclude the former, does not exclude disciplinary technology, but it does dovetail into it, integrate it, modify it to some extent, and above all, use it by sort of infiltrating it, embedding itself in existing disciplinary techniques."[6] Studying these new techniques of power as a

---

Surveiller et punir: Naissance de la prison (*Discipline and Punish: The Birth of the Prison*) first appeared in 1975 because this indicates that questions to do with surveillance were still very much on his mind, but it also suggests that he had brought to a close his thinking about at least some aspects of it. And in fact that is very much the message of his March 17 lecture, which is as much a wrapping-up of work already done (on discipline) as it is an ambitious mapping-out of work that was yet to be done (on biopolitics). While Foucault is quite explicit in stating that discipline and biopolitics are not the same, that does not mean the concept of biopolitics is altogether free from ambiguity—first, he uses the words biopolitics and biopower interchangeably; second, he also uses the word anatomo-politics as an alternate to biopolitics; third, the confusion does not stop there because he also uses the words security and the neologism governmentality as alternates for biopolitics. Perhaps this is it to be expected given that the work Foucault has left us on this topic consists of preliminary sketches, lectures that were never intended to be published in the form we now have them. And he makes it very clear that the work he presents is provisional, as much an attempt to formulate the right questions as it is an attempt to answer those questions. Consequently, there is a great deal of slippage in his terminology, particularly between biopolitics and the later concept of governmentality which he introduces in the academic year of 1977–1978, that is, a year after he introduced biopolitics at the close of 1975–1976 academic year. In this lecture series, published under the title *Security, Population, Territory*, he uses governmentality where previously he had used biopolitics, possibly suggesting that he intended to drop the latter in favour of the former. But in the lectures for the following year, pointedly titled *The Birth of Biopolitics*, it becomes clear that Foucault ultimately came to regard both concepts as equally necessary. One can surmise that he came to consider that the concept of governmentality names his larger project of problematizing the discourse on the way power was conceptualized in political philosophy from the eighteenth century onwards and reserved the concept of biopolitics for the set of administrative technologies whose purpose was to facilitate this transformation. His usage is, however, inconsistent, so this amounts to a kind of secondary revision on the part of his commentators. We may say, then, that both discipline and biopolitics can be subsumed under the general rubric of governmentality, provided we do not lose sight of the key differences between these quite distinct two varieties of power technologies.

[5] Foucault (2003a: 239–240).

[6] Foucault (2003a: 242 emphasis added).

means of better understanding both the nature of power and its operations is the focus of his Collège de France lectures for the next two years. This body of work, despite the slow and until recently ad hoc way it has been published and disseminated, has inspired a tremendous range of applications, across all disciplines in the humanities and social science, but it has tended to give more weight to the 'bio' component of 'biopolitics' than Foucault himself did, thereby (mis)directing our analyses in the direction of smaller and smaller components of the body.

Understood simply as the capture of the human as species by power, which as I'll show is a drastic simplification of Foucault's position, it is not all that surprising that the concept of biopolitics has become so prominent in the twenty-first century. As Roberto Esposito (and many others) has shown we demonstrably live in a biopolitical age. Our physiology has become a site of political contest. Esposito offers several striking examples to make his case, the most perplexing of which is a court case in France in 2000 where it was ruled "that a baby by the name of Nicolas Perruche, who was born with serious genetic lesions, had the right to sue the doctor who had misdiagnosed a case of German measles in the pregnant mother. Against her express wishes, she was prevented from aborting. What appears to be the legally irresolvable object of controversy in the entire incident is attributing to small Nicolas the right *not* to be born."[7] The paradox here, as Esposito explains, is that it is only by virtue of being born that the child gains the very right he wishes to exercise, namely the right never to have existed in the first place.

There is considerable ontological confusion here because it amounts to saying the child who has not been born and who is therefore unknown and literally unknowable to us as a subject nevertheless has the legal right to remain in their state of nonexistence even if they have to come into existence in order to exercise this right. This decision grants a nonbeing the same rights as an actually existing being and in doing so blurs the legal distinction between these hitherto mutually exclusive states of nonbeing and being. Alongside this seemingly absurd ontological situation there is in fact a deeper and quite real political problem, which is that this judgement essentially makes it the duty of the doctor (and perhaps the

[7] Esposito (2008: 3).

parents as well) to eliminate all beings deemed not to be worthy of life because it is a wrong to them if they are born. Esposito goes on to offer several other similarly troubling examples of situations in which the decision regarding who lives and who dies is made along arbitrary lines that take no account of the actual biographies of the people deemed unworthy of life. A glance at any news service today would offer up several comparable examples, making this a very rich seam for contemporary cultural theory.

Foucault himself did not pursue or develop in any great detail this aspect of his work (perhaps because he died before he was able to, as Agamben and others have suggested, though my sense is that this was never the trajectory that interested him). In fact, contrary to the approach of Esposito and others, Foucault explicitly positions biopolitics as a movement away from a direct concern with the actual body, which was the province of discipline, towards something much more abstract, that is, not the empirical dimension of the physiological as such, which is Esposito's concern, but rather the abstract idea of the human species in general as a new type of object for knowledge. The turn towards the biological condition of the body associated with contemporary accounts of biopolitics is, I would argue, a turn in precisely the wrong direction, because what interested Foucault was the way abstractions of the bodily—for example, birth rates, death rates, and so on, which concern bodily attributes and capacities not actual bodies—came to stand in for the body, rather than the body itself, which had become more or less obsolete from the point of view of power.

This is true even of the Nazis, who as Foucault says controlled and regulated the biological aspects of human life more tightly than any other state in history, because what they sought to control were precisely the abstractions of blood quantum (a concept they adopted from US race law), hereditary conditions, birth rates, and so on.[8] In this regard, it is more useful to focus on Foucault's interest in the notion of population as an index of a new type of power than to continue to emphasize the

[8] Foucault (2003a: 259).

biological dimension, which simply leads us to spiral deeper and deeper towards the smallest conceivable biological unit (e.g., DNA, RNA, chromosomes, etc.) as some kind of ultimate reality. Foucault does not take this path. Population is in fact the real focus of his inquiry. He is very explicit on this point. In his opening lecture for his 1978–1979 series of lectures, which are entitled *The Birth of Biopolitics*, he says: "I will try to show how the central core of all the problems I am presently trying to identify is what is called population. Consequently, this is the basis on which something like biopolitics could be formed."[9] To which he adds the important condition that any such analysis of biopolitics would presuppose an understanding of governmental reason, or what he called *governmentality*.[10]

Both discipline and biopolitics are extensions, or better yet apparatuses, of governmentality, which is to say they both presuppose governmentality, so I will focus first on explicating governmentality before I turn to discipline and biopolitics. What is governmentality? Foucault gives several answers to this question, so it's almost impossible to pin it down and define it in a singular way. Foucault himself says that he means three things by the word governmentality—first, it refers to "the ensemble formed by institutions, procedures, analyses and reflections, calculations, and tactics that allow the exercise of this very specific, albeit very complex, power that has population as its target, political economy as its major form of knowledge, and apparatuses of security as its essential technical instrument"; second, it also refers to a general tendency which has led to the pre-eminence of the type of power that we call 'government', which is a demonstrably different type of power to the sovereignty model

[9] Foucault (2008: 22).

[10] It is worth noting here that until all the Collège de France lectures were published more than three decades after they were presented, it was basically on the strength of a single essay culled from those lectures, specifically the lecture given on February 1, 1978, initially titled "On Governmentality", that the concept entered general circulation in cultural theory and rapidly became highly influential, particularly in the social sciences (cultural geography, education, and sociology). As such, although there has been a great deal written about governmentality over the past 40 years, it is really only now that we have a full picture of what it is Foucault was trying to say. The picture we have now is obviously richer than what was previously available and importantly it makes apparent nuances to Foucault's thinking that were hitherto known only to the fortunate few who attended the original lectures.

that preceded it; and third, it refers to a process whereby the state itself is captured by the various techniques of governmentality and effectively 'governmentalized'.[11] This list, which Foucault first offered in 1978, could be greatly expanded by the work he did in subsequent years, so it should be seen as a starting point only. In particular it is missing, or perhaps it would be better to say it underplays, at least three other important domains of reference for governmentality.

As governmentality developed in fits and starts in several places at once across the whole of Europe, leaders of government and theorists of government began to ask three key questions: (1) What should be the concern of government?, (2) What should government know? And, perhaps most importantly, (3) What should be the measure of success for government? Foucault frames the third question in terms of truth. For the past two centuries, or so, government has received its report card from the 'market', which can mean very simply the basic cost or living, or much more complexly the growth of the economy as a whole. If the economy is booming, this is taken to mean government is performing well, that it is doing its job. In recent times, leaders as ideologically opposite as Trump and Obama have both claimed bullish stock markets as evidence of their competence and indeed their legitimacy as leaders. We tend to think of the relation between the market and government in very localized terms, that is, that it is the current set of politicians and bureaucrats employed in government who are doing well, but it goes deeper than that. Foucault means that we accept it as a verdict on the mode of government itself.[12] The economy is thus the 'truth' of government. But there is no reason in principle why this should be so; there are other ways we could evaluate government as various social, environmental, and juridical report cards have tried to do.

Foucault spends comparatively little time on the third question, but the other questions, "What should be the concern of government?" and "What should government know?", command considerable attention. "With what must the state concern itself? For what must the state be responsible? What must it know? What must the state, if not control, at

[11] Foucault (2007: 108–109).

[12] Foucault (2008: 33).

least regulate, or what kind of thing is it whose natural regulations it must respect?"[13] These questions are obviously far-reaching in scope and, as Foucault charts, the scope of these questions constantly broadened from the seventeenth century onwards with, as he says, no structural limit. Indeed, the goal, as Foucault puts it, is to universalize the state such that its mechanisms not only penetrate every aspect of life, but appear to do so by right.[14] Today we would struggle to think of any aspect of our existence that was not subject to governmental scrutiny, regulation, and control. In many cases, though not always, we actually find ourselves welcoming this scrutiny. We have come to think of surveillance as a guarantee of security rather than an invasion of privacy or an impingement of our freedom.

This was demonstrated very clearly by the response to the various lockdown measures imposed around the world to try to contain the spread of COVID-19. In some countries, such as Australia and NZ, people worried that the government hadn't gone far enough in its efforts to contain the spread of the virus. In these countries, people not only willingly complied with the restrictions placed on them, they actively solicited more restrictions and even 'policed' those who didn't comply—Facebook was routinely flooded with images and memes aimed at shaming people for failing to comply with lockdown protocols. In other countries, such as the US and the UK, both the people and the government were much more ambivalent about the need to lock everything down, so the intrusions of government into everyday life were not uniformly welcomed. Many people willingly complied with the various measures—such as the requirements to wear a mask, maintain social distancing, and so on—but others took to the streets with guns (in the US) as a declaration of refusal and in doing so tried to make mask-wearing (among other measures) a liberty issue and not a public safety issue. However, and this is the key point, regardless of whether people thought the government went too far or not far enough there was a general consensus that it was the business of government to do something in response to the pandemic.

[13] Foucault (2007: 350).

[14] Foucault (2003a: 225).

Foucault identifies what he calls governmentality with a shift in political concern (from the point of view of political theory) from the singular focus on the rights and responsibilities of the sovereign to the needs of the population as a whole. How did this shift come about, Foucault inquires, especially given that unlike a war or a volcano eruption one cannot trace everything back to a specific moment in time when this change occurred? Governmentality did not come into being all at once, not even in the ideal and virtual way that Deleuze and Guattari propose for the birth of the state, nor was its birth confined to one place. Its emergence was discontinuous and dispersed, a product not of any deliberate attempts to create something specific whose final shape was known ahead of time, but more the result of a set of moves that resonated and ramified with one another such that they seemed to be caught in a momentum of change pushing government towards a new way of thinking and doing things.

One way this shift can be charted, Foucault suggests, is by tracking the rise and the fall of the family as the principle model of governance in political discourse. From ancient Roman and Greek times, through the European Middle Ages, the art of government, according to the leading political thinkers of the period, is "essentially concerned with answering the question of how to introduce economy—that is to say, the correct way of managing individuals, goods, and wealth within the family".[15] The word economy, derived from the Greek word *oikonomia*, meaning 'household management', is itself an indication of this concern.[16] By the eighteenth century, however, it became clear that this way of thinking had outgrown its usefulness. There were two main reasons for this, both of which are tied to the growth of capitalism: the populations of cities and states had become too large and too diverse to be modelled on the family, indeed they had outgrown the very idea of walled cities; and, the operations of finance, through the rise of new kinds of debt instruments as well as the appearance of derivatives and insurance, had become too sophisticated to be conceived in familial terms.[17] As the market grew in both complexity and wealth so the sovereign's capacity to oversee it declined

[15] Foucault (2000: 207).

[16] Foucault (1985: 154).

[17] Foucault (2000: 215).

because in order to function optimally the market demanded freedoms which were in direct conflict with the sovereignty model.

> One might say this: It is as though power, which used to have sovereignty as its modality or organizing schema, found itself unable to govern the economic and political body of a society that was undergoing both a demographic explosion and industrialization. So much so that that far too many things were escaping the old mechanism of the power of sovereignty, both at the top and the bottom, both at the level of detail and at the mass level.[18]

Power's response, if you will, was the invention of discipline and then biopolitics. The crucial point here is that what power gave with one hand—new freedoms of movement and so on—it took away with the other—by channelling the freedoms towards its own goals. As an aside, one may note here this is exactly the type of dynamic Deleuze and Guattari have in mind for their concepts of deterritorialization and reterritorialization. Mercantilism, the incipient form of capitalism we know today, needed workers who were free to move (i.e., not citizen subjects tied to specific cities, guilds or indentured to master craftsman, and so on), it needed customers with cash to spend rather than goods to barter, and it needed freedom to move goods by land and water in order to reach markets. The walled-city model of sovereignty proselytized by Machiavelli could not accommodate these demands and so over a period of decades its way of conceiving government became obsolete. As Foucault emphasizes, these freedoms when they were granted were not intended to benefit individuals; they were required by the economic system itself, to facilitate the growth of commerce. These new freedoms were not to be had in spite of security mechanisms, but effectively because of them: control is the 'mainspring'.[19] The chained-up subject is much less productive than the properly controlled mobile subject.

The transition to governmentality might have occurred sooner, Foucault speculates, but for the fact that the seventeenth century was particularly tumultuous—climate change in the form of a 'little ice age'

[18] Foucault (2003a: 249).

[19] Foucault (2008: 67).

which caused food shortages in Europe, several major plague outbreaks (e.g., the so-called Great Plague which brought the UK to a standstill in 1666), the thirty years war (1618–1648) which resulted in the deaths of more than 8 million people, financial collapses (most notably Spain when the flow of colonial booty ran dry), peasant uprisings, and so on.[20] A wage-earning apprentice is not necessarily or intrinsically 'more free' than an indentured apprentice. The need to change the way government operated gained traction and urgency in the relative prosperity of the eighteenth century. As the towns and cities of Europe outgrew their fortifications towards the end of the seventeenth century, not only did their populations overspill their rigidly demarcated spaces, they also began to realize that the tightly compressed confines of the old cities (some of which dated to Roman times) inhibited commerce. For this reason, several major cities, including Paris, London, and Rome, began to dismantle their city walls and replace them with thoroughfares, thus enabling better circulation of goods and people. "Broadly speaking, what was at issue in the eighteenth century was the question of the spatial, juridical, administrative, and economic opening up of the town: resituating the town in a space of circulation."[21] It was if, as Foucault says elsewhere, power "found itself unable to govern the economic and political body of a society that was undergoing both a demographic explosion and industrialization. So much so that far too many things were escaping the old mechanism of the power of sovereignty."[22] In the process the character of the city changed, as did the way it was governed.

Whereas previously sovereignty had been exercised within the circumscribed space of the city, now it had to contend with an open space of circulation and exchange. Sovereignty was no longer a matter of territory (much less the family); it had instead become a problem of a mobile population. Traditional conceptions of sovereignty treated the population as a closed set of individuals who collectively embodied the sovereign's power inasmuch as he or she exercised power over them. However,

[20] Foucault (2007: 101). On the 'little ice age', which was precipitated by the genocide of upwards of 50 million native Americans (following European conquest) and the resulting collapse of their agricultural systems, see Bonneuil and Fressoz (2016: 39).

[21] Foucault (2007: 13).

[22] Foucault (2003a: 249).

this form of power can only function within the close confines of a walled city or principality, so as city walls and cloistered principalities broke down new techniques of power were required. This was the problem governmentality answered to. Now government no longer concerned itself with the obedience or productivity of individuals and instead focused on what Marx would later refer to as 'labour power', that is, the abstract form of labour, which relates to hours worked, wages received, surplus value created, but not people. This, I believe, is what Foucault was gesturing towards when he said power captures life itself—the capacity to labour, to be productive, not the biological as such. What really comes through in the lectures, though, is the need to think about governmentality as a form of power for managing an open system: cities defined by circulation rather than hard boundary lines. In these circumstances, productivity replaces obedience as the central concern of government. This does not imply that government relinquished its control over people: it was soon realized that the population can only be the basis of the state's wealth (i.e., productive) if it is properly regulated and as Foucault shows there is no limit to the state's appetite for regulating populations, such that not even the bare fact of being alive is exempt.[23] But it is by no means clear that bare life here means the same thing as it does in Agamben's work, which situates itself in close proximity to Foucault.[24] My sense is that Foucault took the bare fact of living to be the basic condition of possibility for labour, as such it was in the interest of government to be concerned about the welfare of its citizens, as the source of its current and future labour needs.

Biopolitics, as Foucault defines it, is more concerned with the interests of capitalists, particularly their need for the steady supply of well-regulated labour, than it is with the politics of exclusion, which is Agamben's principal concern. This is not to say this issue is unimportant to Foucault. His discussion of racism demonstrates very clearly that it was important to him. But it is to say, perhaps controversially, that his main interest was in the emergence of a new form of discourse relating to the question what counts as good government. This discourse identified two new problems

[23] Foucault (2007: 69).

[24] See Chap. 2.

for the state to consider, neither of which relates to the old order of things: (1) the population is not transparent to the sovereign and (2) it cannot easily be commanded to do anything. "If one says to a population 'do this', there is not only no guarantee that it will do it, but also there is quite simply no guarantee that it can do it."[25] This meant the government of the population could not be thought in terms of either obedience or refusal, compliance or revolt. Since it was seemingly beyond the reach of traditional command and control models of power, for which discipline was developed, the population began to be thought of as a 'natural phenomenon', like weather patterns or ocean currents, with its own similarly invisible and intractable immanent processes, which functioned independently of the behaviour of select individuals and was therefore beyond the reach of discipline.[26] Thus an entirely new apparatus of power was required. Biopolitics solves these two problems, as we will see, by firstly developing the means of rendering the population visible as a population and secondly by developing new security apparatuses that did not require obedience to function effectively. A snapshot of the difference between these ways of approaching government can be seen, conveniently enough given our starting point, in their respective approaches to disease. "At the end of the eighteenth century, it was not epidemics that were the issue [as they had been since the Middle Ages], but something else—what might broadly be called endemics, or in other words, the form, nature, extension, duration, and intensity of the illnesses prevalent in the population."[27] Discipline sought to contain outbreaks of deadly diseases whenever they occurred by locking everything down, while biopolitics sought to reduce susceptibility to all diseases by identifying and addressing the causal factors. The global response to COVID-19 has been a combination of both—lockdown, track and trace, masks, and so on are all disciplinary measures; calculating rates of infection, infection reproduction numbers, and death rates all are biopolitical measures.

---

[25] Foucault (2007: 71).

[26] Foucault (2007: 71). I am tempted to say, this way of thinking about population as a natural phenomenon finds its final form in Deleuze and Guattari, particularly in their dual concept of the molar and the molecular, which is explicitly linked to population. See Deleuze and Guattari (1987: 48–50).

[27] Foucault (2003a: 243).

At this point it may be helpful to take a step back and look at discipline in more detail. Interestingly, according to Foucault it was the need to deal with the ravages of plague which gave "rise to disciplinary projects" in the first place (from the Middle Ages onwards).[28] "Rather than the massive, binary division between one set of people and another [such as one finds in leprosaria and sanatoria], it called for multiple separations, individualizing distributions, an organization in depth of surveillance and control, an intensification and a ramification of power."[29] There are two different modalities of power at work here, according to Foucault: separation (lepers) and segmentation (plague), both of which are still operational today, even in the midst of biopolitics. For example, asylum seekers, when they aren't simply left to drown at sea or die of thirst and exposure in the desert, are bundled into refugee camps and detention centres (the modern-day equivalents of leprosaria) by security forces, where they are treated indifferently—that is, as nameless, stateless, powerless, surplus humanity—and as Foucault puts it, left to their "doom".[30] By contrast, potential terrorists are treated to "meticulous tactical partitioning", just as potential plague victims were, with a similar intent, to separate the infected and infectious from the general population and control their movements and interactions with others. Racial profiling is one of the more well-known tactics deployed by security forces to achieve this end.

In the nineteenth century these two modes of power were combined, such that places of confinement and exclusion—for example, leprosaria, but also prisons, mental asylums, schools, poorhouses, and so on—came to be subject to the same kinds of individualizing processes used to combat plague with the similar aim of 'reforming' the inmate.[31] The key word here is individualizing because, as Foucault stresses, although discipline is applied to individuals, it only does so insofar as those individuals are part of a multiplicity. "The individual is much more a particular way of dividing up the multiplicity for a discipline than the raw material from which

---

[28] Foucault (1977: 198). See also Foucault (2003b: 43–48).

[29] Foucault (1977: 198).

[30] I have drawn here on Reece Jones' *Violent Borders* (2016), which vividly documents the often terrible plight of refugees in the twenty-first century.

[31] Foucault (1977: 199).

it is constructed."[32] This point is crucial because it goes to the heart of how discipline works: it breaks things down to their smallest working parts (in this sense it is an extreme form of Taylorism *avant la lettre*). It "breaks down individuals, places, times, movements, actions, and operations"[33] in order to make them visible and thereby modifiable. The modifications are carried out according to a principle of maximum efficiency, and the optimization of all processes, and on this basis it divides the world into 'normal' and 'abnormal'—the 'normal' is that which is determined as both optimal in terms of its operations and, just as importantly, capable of reproduction. The very idea of 'normal' is thus a product of disciplinary processes rather than an outcome of it.

Emblematic of this outlook and in many ways the pinnacle of this new use of power is the *panopticon*, which thanks to Foucault's scholarly efforts has been lifted from the dusty obscurity of English philosopher Jeremy Bentham's largely unread works of the late eighteenth and early nineteenth centuries and transformed into a widely used concept in the human and social sciences for understanding contemporary forms of power. The word panopticon derives from the Greek word *panoptes* meaning all-seeing. This very neatly sums up the underlying ambition of the panopticon as Bentham conceived it. The panopticon was Bentham's utilitarianist answer to the problem of how to reform prisoners, that is, change their moral character so as to extract labour from them and at the same time reduce the cost of their supervision, and not simply warehouse them as the majority of the world's prisons do today.[34] His hypothesis was that if the surveillance regime could be sufficiently increased in both intensity and scope to the point where it was literally all-seeing then it would induce prisoners to modify their behaviour without further need of coercion and in effect assure "the automatic functioning of power".[35]

The idea was that it should be possible to build a prison in such a way that the prisoners could not escape the all-seeing eyes of the guards and with the help of architects he produced drawings of what he had in mind.

[32] Foucault (2007: 12).

[33] Foucault (2007: 56–57).

[34] See Wacquant (2009).

[35] Foucault (1977: 201).

His proposal was startlingly simple. He imagined a circular building with a hollow centre in which stood an observation tower; all the cells would be arranged in the outer rim of the building, positioned in such a way that they would be completely visible to anyone in the central tower. He further refined this idea by supposing that the prisoners would be backlit to increase their visibility and, crucially, that the central tower be built in such a way that the prisoners would not be able to see whether it was occupied or not. In this way, the cells become "so many cages, so many small theatres, in which each actor is alone, perfectly individualized and constantly visible".[36] The inmate never knows when they are being directly observed so they have to assume they are always being observed; in this way it is not the presence of guards that produces the desired docility but the architecture itself. That, in Bentham's view, was its real beauty.[37] However in spite of the apparent intelligence of its design no panopticon prisons were built in Bentham's lifetime and only a handful have been built since, none of them conforming exactly to his ideas.

We should not dismiss its significance on these grounds, though, because its real importance lay in the way it crystallized a new way of thinking about security. By the same token, it also points to the fact that Foucault was just as interested in the way ideas can have an effect across multiple fronts, without necessarily being realized in the way their originators envisaged. Although the panopticon is generally associated with the prison, the panopticon is not just a "dream building", according to Foucault, it is rather "the diagram of a mechanism of power reduced to its ideal form; its functioning, abstracted from any obstacle, resistance or friction, must be represented as a pure architectural and optical system: it is in fact a figure of political technology that may and must be detached from any specific use."[38] In other words, the panopticon is not necessarily fixed in place, or embodied in a single type of building, whether that is a prison, or a school, but is in fact highly mobile. Indeed, Foucault insists that it should not be identified with a specific building or even a specific

[36] Foucault (1977: 200).
[37] Foucault (1977: 201).
[38] Foucault (1977: 205).

institution—it is, rather, a technology.[39] "The panoptic schema, without disappearing as such or losing any of its properties, was destined to spread throughout the social body; its vocation was to become a generalized function."[40]

Nowadays dispersed CCTV cameras replace the enclosed space and central observation tower model Bentham envisioned, but the principle is the same: an unseen, but all-seeing eye is assumed to be watching us at all times whenever we enter public spaces, especially high-flow spaces like train stations, airports, and shopping malls, and we modify our behaviour accordingly. For Foucault, as it was for Bentham, what is of central importance here is the fact that the panopticon reduces—in some cases eliminates—the need for direct forms of coercion and control. In this sense it can be said to have nullified crime, which is its true aim.[41] It renders the iron fist of power invisible and pervasive—it causes power to pass into the most minute functions of everyday life, such that one can never not be conscious of it. Is this not how one feels moving through an airport? Every step of the way, from the first moment we arrive at the airport, is geared towards individualizing us, placing us, determining if we are a risk or not, and conditioning us to behave in a certain way: our compliance in the face of the indignities of the security checkpoints which strip us naked with their machines is not even requested, it is assumed as the price one must pay to pass through the system. As Primo Levi was instructed on his first day in Auschwitz, there is no 'why?' in the panopticon.[42]

The principal advantages of the panopticon are not hard to grasp. As Foucault puts it, it falls into the category of "It's easy once you've thought of it".[43] Undoubtedly its most important advantage is that it facilitates the supervision of the many by the few, which became essential in Europe in the eighteenth and nineteenth centuries as populations exploded, particularly in the new industrial cities like Manchester, where Engels

[39] Foucault (1977: 215).
[40] Foucault (1977: 207).
[41] Foucault (2008: 256).
[42] Levi (1979: 35).
[43] Foucault (1977: 206).

supervised a factory and Marx spent much of his time. It not only lowers the economic cost of supervision, it also lowers the social cost. Panoptic systems are unobtrusive, so they do not provoke either overt resistance or the overt feeling of oppression in the same way more overt and heavy-handed control systems do. Hidden cameras do not provoke thoughts of resistance in us in the same way that walls beribboned with razor-wire do. Because our fear and anger aren't given an object to respond to, we internalize its presence as a silent order in our heads: the ultimate aim of the panopticon is the docile, self-restraining subject.

The panopticon's effects are continuous, even when its operations are not, as for example when a CCTV camera stops working, because the people who are subject to this type of surveillance are unaware of whether it is working or not. This is why the panopticon was in many ways more important to the growth of industry than the needs of the penal system. Imagine trying to manage a factory on the scale of electronics manufacturer Foxconn's infamous Shenzen plant, which has upwards of 250,000 employees, without sophisticated surveillance systems? It would be impossible. This is why the spread of panopticism went hand in hand with the rise of capitalism.[44] It is therefore ironic, as David Harvey points out, that Foucault is often taken to be anti-Marx, when in fact his analyses can be seen to be both inspired by Marx and adding crucial historical detail to Marx's analyses.[45] However, as I have emphasized already, the panopticon is inherently limited in one crucial respect: it only works in a closed system. Going back to the seventeenth-century expansion of European towns and cities, discipline was not designed for and did not work in these more expansive conditions; its whole approach was unsuited to the demands of the new open city. CCTV does not work in the open city either; it is most effective in the defined spaces of airports, shopping malls, train stations, and so on.[46]

---

[44] Foucault (1977: 218).

[45] Harvey (2010: 149).

[46] As Zuboff reminds us, the actual design of the panopticon, which we owe to Jeremy Bentham's brother Samuel, was inspired by Russian Orthodox churches. "Typically, these structures were built around a central dome from which a portrait of an all-powerful 'Christ Pantokrator' stared down at the congregation and, by implication, all humanity. There was to be no exit from this line of sight." Zuboff (2019: 470–471).

What then is the difference between discipline and biopolitics? First of all, in contrast to discipline, which focuses on bodies, biopolitics is "is applied not to man-as-body but to the living man, the man-as-living-being; ultimately, if you like, to man-as-species."[47] Foucault clarifies what he means here by adding "I would say that discipline tries to rule a multiplicity of men to the extent that their multiplicity can and must be dissolved into individual bodies that can be kept under surveillance, trained, used, and, if need be, punished".[48] The principle process of discipline is individualization, which in turn is followed by four main types of operation: selection, normalization, hierarchization, and centralization.[49] By contrast, biopolitics is not interested in the individual as an individual, but only as a component of an essentially indissoluble multiplicity, namely a population that it approaches from the perspective of large numbers (i.e., it looks for tendencies and trajectories across time rather than isolated events). "Biopolitics deals with the population, with the population as political problem, as a problem that is at once scientific and political, as a biological problem and as power's problem."[50] More importantly, though, it constitutes a very different approach to government. Conveniently, we can return here to the example of the pandemic I mentioned above: Foucault argues that towards the end of the eighteenth century, the focus of what we might term governmentalized medicine shifted away from a concern with infectious disease epidemics which still killed the majority of people to a focus on endemic illnesses. "These were illnesses that were difficult to eradicate and that were not regarded as epidemics that caused more frequent deaths, but as permanent factors which—and that is how they were dealt with—sapped the population's strength, shortened the working week, wasted energy, and cost money, both because they led to a fall in production and because treating them was expensive. In a word, illness as a phenomena affecting a population."[51]

[47] Foucault (2003a: 242).
[48] Foucault (2003a: 242).
[49] Foucault (2003a: 181).
[50] Foucault (2003a: 245).
[51] Foucault (2003a: 243–244).

The practical aim of biopolitics is to control the random by being better able to predict the likelihood of specific eventualities, whether that's the incidence of disease, crime, or terrorism, the principle is the same.[52] And its principle weapon, at least in its first iteration, is statistics, which etymologically means literally 'science of the state'.[53] This approach yields a new object, the population, which it tries to grasp in motion so to speak with a view to seeing what does not vary, its constants in other words. Biopolitics investigates social phenomena "that [over time] are aleatory and unpredictable when taken in themselves or individually, but which, at the collective level [i.e., the level at which the law of large numbers applies], displays constants that are easy, or at least possible to establish".[54] This leads to the creation of very different types of control mechanisms from those deployed by discipline, such as "forecasts, statistical estimates, and overall measures", not with a view towards modifying the situation for a particular individual (as discipline does), but rather "to intervene at the level of their generality".[55]

Today this way of thinking about government is thoroughly institutionalized, as can be seen readily enough in the field of medicine: there are entire fields devoted to public health, population health, epidemiology, and so on, that think almost exclusively in terms of identifying and mitigating health risk factors, all with a view to maximizing the number of healthy days the population as a whole enjoys and conversely minimizing the number of unhealthy days and hospitalizations required by the population. As Foucault says, the goal is precisely to "optimize a state of life".[56] But, as Foucault reiterates, biopolitics is not concerned with the individual; it works at the collective level, at the level of the greatest generality, which is the level of life itself. Biopolitics takes hold of "the biological processes of man-as-species [...] ensuring that they are not disciplined, but regularized".[57] Biopolitics concerns death rates, not

[52] Foucault (2003a: 259).
[53] Foucault (2007: 101).
[54] Foucault (2003a: 246).
[55] Foucault (2003a: 246).
[56] Foucault (2003a: 246).
[57] Foucault (2003a: 246–247).

deaths; life expectancies, not lives; births rates, not births; employment rates, not employment; jobs growth, not jobs; and so on.

The explicit emphasis on optimizing life points towards an aspect of biopolitics that is rarely discussed, which is the fact that it is an inherently ambivalent concept. Although Foucault presents biopolitics in such a way that its emergence seems to have been largely negative, at least from the point of view of individual freedoms, and that is certainly how it has been cast by Agamben and other scholars following in his footsteps, because it is a new form of power, the reality is slightly more complex. As the COVID-19 pandemic made abundantly apparent, we all wished that governments had done more to contain the outbreaks when they occurred. COVID-19 viewed as an isolated, unexpected outbreak of a deadly disease would not be a matter for biopolitics. However, COVID-19 was neither isolated nor unexpected—pandemics are regular occurrences whose effects can be calculated, their trajectories predicted and mapped.[58] Most wealthy nations have pandemic response plans in place, for this reason, ready to be mobilized as they may be required, and we generally regard this as an example of good government at work. Indeed we have come to expect it as a minimum form of government.

Trump's decision in 2018 to get rid of the pandemic response team Obama set up is seen by many (albeit from the perspective of 2020) as an unambiguous example of bad government. Similarly, his decision to withdraw US funding from the WHO in the early days of the pandemic was widely seen as bad government. His anti-mask stance, his push to send kids back to schools and open businesses while the disease was still rampant are seen by most people as an abrogation of the very idea of government. When pandemics occur we want biopolitics, so much so that we interpret its absence as a failure of government and in cases like COVID-19 which caused considerable loss of life we see it as a form of structural violence. Many people feel the same way about childhood vaccination programmes, old-aged pensions, environmental regulations protecting air and water quality, bank deposit guarantees, and so on, all of which are examples of biopolitics at work according to Foucault.[59] We

[58] Davis (2005).

[59] Foucault (2003a: 244).

have come to expect that the government ought to be responsible for providing some or all of these things, as such we do not automatically regard them in a negative light as a loss of personal freedom.[60] This brings us to the other point of difference between discipline and biopolitics, namely the sinister fact that Trump's decision to reduce his government's capacity to deal with a major pandemic is also a decision to let people die. And that is precisely what happened in 2020 when COVID-19 struck; it killed more people in the US than anywhere else in the world despite the fact the US is the richest country in the world, with vast resources at its disposal.

If we were shocked by Trump's callousness it is not only because of his blatant disregard for human life but also because it flies in the face of our understanding of what we see as the central mission of government today, which is *to make live*. Biopolitics reveals its truth as a new form of power in the emphasis it places on optimizing life. This stands in stark opposition to the model of power Foucault refers to as sovereignty, which defined itself in relation to death, or more precisely the right of the sovereign to take life. According to "the classical theory of sovereignty, the right of life and death was one of sovereignty's basic attributes".[61] In practice this amounts to saying the sovereign has the right to kill: "it is at the moment when the sovereign can kill that he exercises his right over life".[62] This right was not generally regarded as absolute or unconditional—it could not be exercised on a whim in other words; rather, it could only be exercised when external enemies attacked or sought to overthrow the sovereign. Then, in order to preserve both themselves and their crown the sovereign could invoke a 'state of exception' and exercise their right to kill. A constant reminder of the sovereign's right to kill is to be found in the image of 'Lady Justice', which adorns legal institutions all over the world—she is blindfolded and carries scales in one hand and a sword in the other. Justice is blind, fair, and lethal.[63] However, as governmentality

[60] Neoliberalism's strategy for talking us into thinking that these things are negatives takes precisely this path: vaccinations, pensions, environmental regulations, bank regulations, and so on are all presented as infringements of individual freedom that we would be better off without.

[61] Foucault (2003a: 240).

[62] Foucault (2003a: 240).

[63] Foucault (1978: 135–136).

supplanted sovereignty this way of thinking about power found itself caught in a trap of its own making: "How can a power such as this kill, if it is true that its basic function is to improve life, to prolong its duration, to improve its chances, to avoid accidents, and to compensate for failings?"[64] Foucault's answer to this question is both simple and complex: racism.

> What in fact is racism? It is primarily a way of introducing a break into the domain of life that is under power's control: the break between what must live and what must die. The appearance within the biological continuum of the human race of races, the distinction among races, the hierarchy of races, the fact that certain races are described as good and that others, in contrast, are described as inferior: all of this is a way of fragmenting the field of the biological that power controls. [...] That is the first function of racism: to fragment, to create caesuras within the biological continuum addressed by biopower.[65]

This much is probably obvious, but it has a second function which is less obvious, because it is a transformation of what we might call the traditional way of thinking about war, which says if you want to live, then the other must die. According to Foucault, "racism makes it possible to establish a relationship between my life and the death of the other that is not a military or warlike relationship of confrontation, but a biological-type relationship: 'The inferior species must die out, the more abnormal individuals are eliminated, the fewer degenerates there will be in the species as a whole, and the more I—as species rather than individual—can live, the stronger I will be, the more vigorous I will be. I will be able to proliferate."[66] The death of the other does not guarantee my security, as it would in a war; rather it makes me healthier. But this way of conceiving the other is no longer political, it is biological, and in that sense the very nature of power itself has changed its complexion because now "the enemies who have to be done away with are not adversaries in the political sense of the term; they are threats, either external or internal, to the

[64] Foucault (2003a: 254).
[65] Foucault (2003a: 254–255).
[66] Foucault (2003a: 255).

population and for the population."[67] To put it very crudely, in a political system that is premised on optimizing the lives of an entire population, the only way it can then turn around and justify killing people is by coding those people as harmful to the health of the population as a whole. Racism is, in this sense, politically necessary, because it "alone can justify the murderous function of the state".[68] Importantly, Foucault goes onto say that racism first developed with settler-colonialism. "From this point onward, war is about two things: it is not simply a matter of destroying a political adversary, but of destroying the enemy race, of destroying that [sort] of biological threat that those people represent to our race."[69] War becomes, then, a means of regeneration, purification, and so on, for the nation waging it. This murderous mode of racism is quite different then from either ordinary forms of racism that manifest as contempt or ideologically loaded discourse: it is, as Foucault puts it, a "technology of power".[70] It is an intrinsic part of the state—we could say even that it is governmentalized.

The treatment of asylum seekers provides a clear-cut example of what Foucault means here. This is particularly true of countries like Australia which respond hysterically to the influx of several thousand asylum seekers as though it were a human tidal wave (the word inundation is frequently used in this context), which even if true would scarcely matter in a country so sparsely populated. The argument against allowing asylum seekers into the country usually takes two forms, both of which are coded ways of speaking about the 'health' of the nation. Most obviously, it is spoken about as a problem for the nation's cultural identity, as though to say the influx of others will somehow 'dilute' the existing culture or in other words introduce impurities. The idea that the culture was somehow pure to begin with is of course a fantasy, but that does not mean it doesn't feel true to the people who espouse it. Not only that, it deliberately forgets that as a settler-colonial country the majority of the population are in fact migrants, many of them quite recent, so its culture is in fact a

[67] Foucault (2003a: 256).
[68] Foucault (2003a: 256).
[69] Foucault (2003a: 257).
[70] Foucault (2003a: 258).

hybrid form born of decades of mixing of cultural forms. These kinds of complaints are often dismissed as 'culture wars' and intelligent people tend not to buy into them, but the point I want to make is that they precisely follow Foucault's model that of racism manifesting as the desire to protect the health of the race.

The second way this logic manifests itself is much less obvious and therefore much more pernicious. Asylum seekers are portrayed as a 'drain' on the economy (the prevalence of these hydrological metaphors would make an interesting study in its own right), which for a rich country like Australia is simply untrue, a point that is amply proven by the fact Australia spends more to detain asylum seekers in hellish offshore prisons than it would cost to treat them humanly and provide them with social housing in Australia. Not only that, Australia has a long history of taking in refugees and asylum seekers (more than 2 million following World War 2), many of whom now oppose the entry into the country of any other refugees or asylum seekers, a fact that is conveniently forgotten. The economy is portrayed as a body whose vitality depends on it being managed carefully and not frittered away on foreign others. Instead of thinking of helping others makes Australia a stronger more humane nation, helping others is depicted as risking (unacceptably) the health of the nation. In this way the inherent racism of the anti-asylum seeker stance is obscured and the position made to seem somehow 'rational' because it appears to be just a matter of good economics. The logic underpinning it, though, is very clear.

This formulation may also help us to explain one of the more curious phenomena of the COVID-19 outbreak—the appearance of anti-maskers. Social media bore witness to countless episodes of people being filmed refusing to wear face masks in public and creating a scene when other people demanded that they in fact wear a mask. Typically they would refuse to wear a mask on the grounds that it infringed their freedom, variously as Americans, or Australians, or whatever, but also even more astonishingly as humans. It is hard to conceive how anyone could construct a view that wearing a mask was an impingement on freedom, particularly when the only point of wearing the mask is to protect their health and the health of others. But obviously in thinking this we are already thinking in terms that are recognizable as an instance of governmentality at work, it shows an expectation that the government should intervene to modify the behaviour of individuals for the greater good. As many people pointed out on social media, it's not as though we don't already wear seatbelts in cars, helmets when riding a bike, and so on, which could also be construed as impinging our freedom, and do so largely without a whimper.

So why did mask-wearing provoke such hostility and outrage? One answer to this question is, I believe, race. Although social media is anything but a reliable source, it appears that it was predominantly, if not exclusively white people who took this stance, and in most cases it was poor white people. If, as Foucault suggests, the racial other is construed as a health hazard, then perhaps the reason the requirement to wear a mask is so triggering is that it positions the wearer in the category of the racial other, as the health hazard, and that is simply intolerable to some people. To the white racist it amounts to a negation of their whiteness, which may in fact be the only shred of privilege they feel they have in the world, so it feels like a matter of survival to them. Their unprovoked hostility towards anyone suggesting they ought to wear a mask for the sake of everyone's health, and not just because it's a requirement, is in this sense a symptom of just how negated the mask makes them feel.

References

Bonneuil, C., and J.-B. Fressoz. 2016. *The Shock of the Anthropocene: The Earth, History and Us*. Trans. D. Fernbach. London: Verso.

Davis, M. 2005. *The Monster at Our Door: The Global Threat of Avian Flu*. New York, United States: New Press.

Deleuze, G., and F. Guattari. 1987. *A Thousand Plateaus: Capitalism and Schizophrenia*. Trans. B. Massumi. Minneapolis: University of Minnesota Press.

Esposito, R. 2008. *Bíos: Biopolitics and Philosophy*. Trans. T. Campbell. Minneapolis: University of Minnesota Press.

Foucault, M. 1977. *Discipline and Punish: The Birth of the Prison*. Trans. A. Sheridan. London: Penguin.

———. 1978. *The History of Sexuality: An Introduction*. Trans. R. Hurley. London: Penguin.
———. 1985. *The Use of Pleasure: The History of Sexuality Volume 2*. Trans. R. Hurley. London: Penguin.
———. 2000. *Essential Works of Foucault 1954–1984 Volume 3*. Trans. R. Hurley et al. New York: The New Press.
———. 2003a. *Society Must be Defended: Lectures at the Collège de France 1975–1976*. Trans. D. Macey. New York: Picador.
———. 2003b. *Abnormal: Lectures at the Collège de France 1974–1975*. Trans. G. Burchell. New York: Picador.
———. 2007. *Security, Territory, Population: Lectures at the Collège de France 1977–1978*. Trans. G. Burchell. New York: Palgrave.
———. 2008. *The Birth of Biopolitics: Lectures at the Collège de France 1978–1979*. Trans. G. Burchell. New York: Palgrave.
Harvey, D. 2010. *A Companion to Marx's Capital*. London: Verso.
Jones, R. 2016. *Violent Borders: Refugees and the Right to Move*. London: Verso.
Levi, P. 1979. *If This Is a Man*. Trans. S. Woolf. London: Abacus.
Wacquant, L. 2009. *Punishing the Poor: The Neoliberal Government of Social Insecurity*. Durham, NC: Duke University Press.
Zuboff, S. 2019. *The Age of Surveillance Capitalism: The Fight for a Human Future at the New Frontier of Power*. London: Profile Books.

2

The Market Lives on Death: The Endocolonizing Logic of the Fascist Moment

Samir Gandesha

Over the past several years, if not decades, ghosts of fascism have escaped their twentieth-century crypts and come to haunt our present. With the global COVID-19 pandemic, however, we face the prospect of our "Reichstag Fire" moment. This was an arson attack on the German legislature exactly four weeks after Adolf Hitler was sworn in as Chancellor, allegedly carried out by Marinus van der Lubbe, a Dutch Council Communist. The Nazis immediately claimed that the fire was the result of a communist plot and it became the pretext for their seizure of power (*Machtergreifung*) and total co-ordination of the state (*Gleichschaltung*). As was recently pointed out by the *Economist*, close to a dozen states, from Azerbaijan to Togo, have already used the pandemic to arrogate more power to themselves. Indeed, this development has been particularly visible in Washington, Budapest, and Delhi.

Donald J. Trump has claimed "total authority" for the Oval Office in opposition to state governors who had sought to loosen lockdown

S. Gandesha (✉)
Institute for the Humanities, Simon Fraser University, Vancouver, BC, Canada
e-mail: insthum@sfu.ca

S. S. Das, A. R. Pratihar (eds.), *Technology, Urban Space and the Networked Community*, https://doi.org/10.1007/978-3-030-88809-1_2

measures earlier. While he quickly backtracked on this claim, he has, nonetheless, more recently called upon his supporters in blue states to resist lockdown measures and "liberate" themselves from the authority of Democratic governors in an effort to get the wheels of the economy turning again, and has closed the US's borders and suspended immigration for sixty days. The implicit identity of the health of the bodies of individual (White) Americans with that of the US body politic is clear.

Viktor Orbán, Hungary's president, having previously curtailed the autonomy of the courts, has indefinitely suspended the legislative branch of government, eliminating, in the process, the key liberal-democratic principle of institutional limits on executive authority—he now rules by decree. Orbán has consistently, over the years, attacked George Soros whom he has taken as the metonym of the baleful "globalist," which is to say, Jewish, influence on Hungarian politics. Former Canadian Prime Minister, Stephen Harper, is one of Orbán's many admirers.

The RSS in India—the quasi-fascist Hindu nationalist (Hindutva) force behind Modi—has, in a classically fascistic move, characterized its Islamic "enemy" as the abject carrier of the COVID-19 virus. The hashtags "CoronaJihad" and "BioJihad" have proliferated via Twitter, as Jason Stanley and Federico Finchelstein have recently indicated via Indian journalist Rana Ayyub, just as the Nazis had used typhus as the pretext for excluding Jews, isolating them in ghettos and ultimately murdering them. The targeting of Muslims comes in the aftermath, of course, of the unconstitutional annexation of Kashmir and changes to the Citizenship Act that explicitly and unapologetically discriminate against this oppressed and reviled minority community.

Fascism?

Be this all as it may, one must always, nevertheless, be careful when using the word "fascism." The term is often used so indiscriminately—especially on the Left—to vilify one's political opponents that it is in continual danger of losing all meaning. In what sense, then, can we say that what we are witnessing throughout the globe is the re-emergence of fascism? Writing in the pages of the *New Left Review* two years ago, Dylan Riley has argued trenchantly that if we compare twentieth-century fascism with contemporary authoritarians such as Trump across four axes: geopolitical dynamics; the relation between class and nation; developments within civil society; and political parties, there is no persuasive evidence that what we are confronted with today is anything approaching fascism. And, indeed, according to Slavoj Žižek's influential gloss on Walter Benjamin, the authoritarianism that we see around us today does not arise in response to what could reasonably be called a "failed revolution." Of course, there were the Arab Spring and the Occupy movements, but they did not come remotely close to challenging the domination of capital.

However, as Samir Amin has perceptively argued, fascism does not have to entirely conform to the twentieth-century mould and may be simply understood as being comprising two essential elements. The first is that it is the response to the crisis of capitalism. The second is that it constitutes a categorical rejection of "democracy" by way of an appeal to collective identities—often condensed in the figure of a "strong" leader—tied to a notion of the "people." Yet, two refinements ought to be made to Amin's definition.

The first such refinement is that the very notion of crisis needs to be rethought. Under neoliberalism, crisis is no longer to be regarded as discrete and cyclical but rather as continuous and enduring. It is not an event but a syndrome or a condition; to use a medical metaphor, crisis is no longer "acute" but "chronic." This means, of course, that fascism is always something of a haunting presence within a neoliberalism that is, one could say, coextensive with a deep and abiding fissure within the social order. Once precisely touted as the antidote to authoritarianism (see the writings of Hayek and Foucault), neoliberalism deepens and exacerbates authoritarian tendencies that are coextensive with capitalism itself. That, at the end of the day, capitalism, now faced with crushing inequalities (see Piketty's flawed but nonetheless useful *Capital and Ideology*), will preserve itself by any means necessary. As Theodor W. Adorno once argued, the real threat of fascism comes from within not from outside of capitalist or liberal democracy.

The pandemic undoubtedly overdetermines neoliberalism's endemic crisis and the word crisis, it is important to remember, derives from the Greek *krisis* (decision) and *krinein* (to decide). In late Middle English, the word comes to mean the turning point of a malady, that *decisive* point at which time the condition of the patient manifestly *improves* or *deteriorates*. In other words, if the crisis of our neoliberal social order—greatly over-determined and exacerbated by the COVID-19 pandemic—is chronic rather than acute, then it is a time at which the figure of the sovereign, the entity that *decides* on the exception (Carl Schmitt), comes to cast a particularly long and dark shadow over our times.

This is exemplified by the way in which, as discussed above, in the US, Hungary, and India, the pandemic, less of a definite *event* than an amorphous *syndrome*, plays the role of the "Reichstag fire" in consolidating sovereign power. As it has been widely observed, the pandemic brings into visibility the deep-seated precarity constitutive of the neoliberal order—one that only the very wealthy can seemingly escape.

The second refinement is that fascism is not a categorical rejection of democracy per se but rather a rejection of its *liberal* form. As Vladimir Putin recently mused, perhaps liberal democracy is obsolete. Yet, like leaders of fascist movements of the twentieth century, Putin makes an appeal of a certain sort to the idea of democracy (Rousseau's "general will as opposed to what he calls the "will of all"). He does so by claiming to embody the will of the *demos* people or *Volk*, and this is what makes such claims especially dangerous today. There is, in other words, considerable overlap between twentieth-century fascism, on the one hand, and contemporary forms of right-wing or authoritarian populism—which are often correctly described as "neo-fascist" or "post-fascist," as Enzo traverso recently indicated—on the other.

An important difference between twentieth- and twenty-first-century forms of fascism is that while the former in Germany, under the pretext of the post-Reichstag Fire emergency, abolished the right of assembly, freedom of the press, and ultimately, elections by suspending the Weimar Constitution, right-populists today are committed, at least nominally and for the time being, to contesting elections, although they are quite happy to dispense with many of its corollaries such as the rule of law, respect for minority rights, the division of powers, engagement in gerrymandering, voter suppression, and so on. In fact, they are mobilizing divisions so effectively that they are winning elections and maintaining popular support for the time being, particularly in Hungary and India.

More than that, as we saw on January 6, 2021, Trump, with help from his associates such as Rudi Giuliani, incited his followers to attack the Capitol on the basis of his claim that the election had been stolen by Joe Biden, Kamala Harris, and the Democratic Party. In the violent assault on what was a literal and figurative attack on the hallowed institutions of US democracy, scores of people were injured and several died both on the day and in the days that followed. Later it was revealed that several Republican law makers were complicit in the planning of the attack, which included a plan to execute Democratic congressmen and women. Trump's second impeachment trial for the incitement of the insurrection was decided largely according to party lines which means, among other things, that the commitment of the Republican Party to the liberal-democratic framework of US democracy as outlined in the country's constitution is, at best, ambivalent. Such ambivalence is reflected in countries such as India, Brazil, and Hungary, among others (Fig. 2.1).

A vitally important difference between the fascism of the twentieth century and that of the present century is how each of them conceive of time. To be sure, Hitler's dream of a "1000-year Reich" was spatially oriented, insofar as it was based on the securing of *Lebensraum* to the east for the German *Volk*. However, what was more important than *space* in Nazi

Fig. 2.1 Fascism? Lawfare Blog, Jan 7, 2021

thinking was *time*, insofar as fascism was, in its own perverse way, "utopian" and "revolutionary" oriented to a bright new future for the "Aryan race." In *Being and Time*, card-carrying member of the Nazi Party, Martin Heidegger, elevated the temporal modality of the future over both the past and the present. The future would be secured by retrieving the forgotten experiences at the origin of the Ancient Greek understanding of Being.

Present-day fascism, in contrast, takes refuge in the past as such: in a supposedly "great" America before the Civil Rights Act (if not before the Civil War); in an authentic homeland of the Magyars in Hungary; and in a purified India for Hindus (Hindustan). In other words, contemporary fascism makes little or no claim on the future in an era of its *ecologically planned obsolescence*. In this, it is, as Aimé Céasire had already pointed out in *Discourse on Colonialism*, a form of European colonialism applied to Europe itself—*endocolonialism*, as it were. Today, as we have seen in the Greek case, its weapons are German banks rather than tanks. As I have written about elsewhere, we see this, as well, in the willingness of that model of kumbaya "liberal multiculturalism"—the Canadian state—to deploy the logic of the exception to permit on-going large energy infrastructure projects (hydro, LNG, and bitumen) under conditions of a COVID-19 lockdown. Recalling the weaponization of disease in the earliest days of contact between Indigene and Colonizer, this puts already vulnerable Indigenous communities at serious risk of a health catastrophe. As the meme goes "Genocide is not an essential service!" The same logic can be discerned in the Modi Government's resource extraction agenda driving the war on India's tribal peoples (Adivasis) in Chhattisgarh, not to mention in Jair Bolsonaro's iron-fisted developmental programme in the Amazon basin.

Spectres of fascism loom, then, as a response to the chronic financial and ecological crisis of capitalism. Fascism in the twentieth century offered, in part, a solution to the economic slump via an acceleration of the extraction of absolute and relative surplus value from living labour by smashing the revolutionary Left, independent trade unions, and other working-class institutions. This was, indeed, the original meaning of Mussolini's (and Gentile's) idea of fascism based on the image of what was called, in Latin, *fasces*, a bundle of rods and protruding axe blade symbolizing the penal powers of the Roman state wielded by the magistrate. Fascism entailed, then, the binding together of the rods of the state, capital, and labour. It is, perhaps, telling that both the US and French Republics adopted and maintained this proto-fascistic Roman symbolism through the eighteenth and nineteenth centuries.

In contrast to its anti-human twentieth-century form, contemporary "post-human" fascism centres on a deepening of resource extraction on the very precipice of massive deskilling of labour, and widespread automation and employment of robotics, machine learning, and artificial intelligence to wit—the prospective obsolescence of humanity itself. Such a logic entails what, in *Critique of Black Reason*, Achille Mbembe calls the "becoming Black of the world," the creation of "abandoned subjects":

> There are no more workers as such. There are only laboring nomads. If yesterday's drama of the subject was exploitation by capital, the tragedy of the multitude today is that they are unable to be exploited at all. They are abandoned subjects, relegated to the role of a "superfluous humanity."

This superfluousness now becomes clear as governments, by omission or commission amidst the pandemic, put members of society deemed surplus as well as workers, particularly people of colour, at grave risk of contracting or even dying from the virus (a recent UCSF study conducted in San Francisco's Mission District showed that 95% of positive cases were Latinx). Of course, it could be argued that human labour has never appeared more "essential" than in this historical moment. Yet, states are also showing themselves quite willing to put essential workers at such an extreme risk as to even die *en masse* for want of PPE, for example. MTA conductor and writer, Sujatha Gidla, reports her co-workers as saying "we are not essential, we are sacrificial" (*New York Times*, 6 May).

In his depiction of the aftermath of catastrophe, possibly nuclear war, in *Endgame*, Samuel Beckett presents the destruction of nature as taking a specific spatial configuration in which the dialectic of time, itself, has seemingly come to a standstill. He shows, in coldly unsentimental though often humorously ribald terms, the obsolescence of human beings, reduced as they are to pure existence, and subordinated to the inscrutable machinations of geopolitical forces beyond their understanding and control. The necessary supplement to *Endgame*, according to Stanley Cavell, is Kubric's Cold War masterpiece *Dr. Strangelove*.

Beckett depicts the parents of his anti-hero, Hamm, as literally reduced to a form of societal refuse, having been confined to garbage bins—perhaps signifying for us today, all-too painfully, the perilous state of nursing homes—warehouse-coffins for human beings poised somewhere between life and death, waiting for an end to the excruciating game of waiting. They wax nostalgic ("Ah, the good old days," sighs Nell) about the days when they were provided with sand, rather than sawdust, in their metallic cloisters, a signifier of happier times spent on the beach rather than of a nature that is now "corpsed." The catastrophe of the present and its relation to the recent past forms a continuum of the same unfolding disaster Walter Benjamin writes about in his final text "On the Concept of History" before his desperate flight from the Nazis and consequent suicide in Port Bou. Today, governments seem prepared to sacrifice the elderly, the infirm, poor, indigent, Black, and brown to the iron laws of the market. Republican Lieutenant Governor of Texas, Dan Patrick, recently suggested that grandparents might consider *sacrificing* their health and lives for their grandchildren, which is to say, for the health of the economy. But this logic is nothing new. It was previously discernible in each press release from myriad corporate head offices of massive downsizings producing an immediate, dramatic uptick in their share prices. The market lives on death (Fig. 2.2).

If we take as our definition the classic account of fascism as that revolutionary mass movement comprising an alliance between industrial capital and the petite bourgeoisie ranged against the working class and its political organizations, in the context of imperialist rivalries and discrete capitalist crises of overproduction, then it is far from clear that what we face today can be described as "fascism." After the defeat of organized labour, there is precious little resistance to dead labour's machinic extraction of surplus value from living. Such a defeat clears the way for redoubled colonization and endocolonization, racism, militarism, and, ultimately, war.

The endocolonizing logic of contemporary fascism becomes particularly manifest in the context of the contemporary cityscape. Here we can

Fig. 2.2 Nagg and Nell in Beckett's *Endgame*, Shimer College, Chicago, 2009. (Collected from wikimedi.org)

extrapolate the rationality of the post-colonial periphery in the institution of policing, which, of course now takes on new meaning, with the election of the Joe Biden and his vice-president former San Francisco DA and California Attorney General, Senator Kamala Harris. If behind every fascism can be discerned a failed revolution, then it is possible to see the contours of the contemporary fascism moment not simply in the rise of the hard right turn of the Republican Party with Newt Gingrich and the Tea Party, but in Nixon's campaign against the protest movement, the Black Panther Party, in particular. Far from the Biden-Harris triumph as representing a decisive defeat of a quasi-fascistic Trumpism, what it does is reinforce the failure of the very revolution that Harris's parents sought to participate in. In this sense, Harris is the uncanny double of Angela Davis, whom another Bay Area prosecutor sought to execute on trumped up charges as was so well documented in Shola Lynch's moving 2012 film, *Free Angela and All Political Prisoners*. A glance at Ousmane Sembène's classic film *Borom Sarret* (The Wagoner) draws out, proleptically, some of the features of the endocolonial cityscape today. After a brief discussion of the film, I fast-forward to a viral YouTube video of African-American author and activist, Kimberley Jones's critique of the concept of the "looter" in the context of #BLM protests.

Policing the Post-colony

Widely regarded as the first film made in Black Africa, *Borom Sarret* (The Wagoner) by Ousmane Sembène provides a profound glimpse of immediate post-colonial reality (Fig. 2.3). Made in 1963 upon the auteur's return from learning his craft at the Gorkii Studios in Moscow, it portrays the unfolding of a day in the life of a cart driver in Dakar, Senegal.

Fig. 2.3 Policing the post-colony: stills (collected) from *Borom Sarret* (The Wagoner) Dir. Ousmane Sembène

Its formal minimalism enables *Borom Sarret* to reveal several layers of complexity. In the economical space of approximately 18 minutes, it discloses the structural violence established and consolidated through colonial class and gender relationships that live on, uncannily, in the post-independence period. It is a vivid and crystalline cinematic depiction of what Frantz Fanon had called just two years earlier in *Wretched of the Earth*, the "pitfalls of national consciousness" and the way in which precisely such an imaginary served to mask the real, which is to say ruthlessly exploitative relationships among citizens of newly "liberated" states. It provokes suspicion of the now ubiquitous idea, at least in the global north, that the abstraction of racial identification alone could ever be organizing principle of solidarity and therefore politics.

We follow the driver, and are privy to his interior monologue delivered by Sembène himself, while he transports a series of passengers and materials to their various destinations. The cart driver considers the exertions of an unemployed man futile and irritating; he is coldly unsympathetic to his plight. He is accosted by a severely crippled yet reasonably affable beggar who asks for money but and is even less solicitous and ignores him: "there are so many of them, they are like flies." Yet, the driver is more than happy to pay the well-fed and well-dressed griot or folk singer, who builds up the driver's ego ideal by his ingratiating and obsequious praise of the warrior-identity of his ancestors.

Then there's the solemn father whom the driver transports with the corpse of his infant child to the cemetery only to be turned away because his papers are not in order; he is, we learn, a "foreigner." The artificial borders of the "nation-state" constructed ex nihilo by the colonial powers continue to enact their violence, unremittingly, on the most vulnerable. The driver carefully places the corpse of the child on the ground and drives way, leaving the bereft father to suffer alone.

The narrative begins to tighten with the approach of a well-dressed and apparently wealthy African man who wishes to be taken to the formerly French quarter of Dakar—the Plateau; here, cart drivers require special permits. The man is moving to the Plateau, he tells the driver. The camera pans in the direction of the former European quarter to reveal a shockingly different cityscape. As the soundtrack shifts from the syncopated rhythms and xalam (lute) of traditional Senegalese music to

Fig. 2.4 Policing the post-colony: stills (collected) from *Borom Sarret* (The Wagoner) Dir. Ousmane Sembène

eighteenth-century European classical music, the sand and rock give way to paved streets, the horse-drawn carts to orderly modern automobile traffic. In a few short miles, we traverse centuries (Fig. 2.4).

As soon as the driver nervously enters the Plateau, he is immediately confronted by a scowling police officer who promptly issues him a fine and confiscates his cart. As he is writing the ticket, the officer steps on the wagoner's medal, most likely for the driver's service in the French army. Meanwhile, the wealthy passenger absconds in an awaiting car. In this single gesture, the continuity of the corruptions of Empire is laid bare. Racial solidarity is revealed for the myth that it is. The police are there to protect the wealthy Blacks from poor Blacks, whose labour power is nonetheless required for the production of wealth; the inclusion of the worker is premised on their spatial exclusion. They are what Jacques Rancière calls "the part that has no part."

The driver returns home with his horse, devastated and bewildered. His wife rise, matter-of-factly gives him their infant child to look after, and promises that they would have food that evening and leaves. According to the Director of NYU's Institute of Afro-American Affairs, Manthia Diawara, the common interpretation—consistent with themes in Sembène's other films—is that she is off to participate in sex work and this was not to be disparaged but accepted as a legitimate form of labour; sex workers were to be accepted as proletarians and neither stigmatized nor condemned, as they were, of course, by the imams.

Today in the midst of the global uprising, amidst the COVID-19 pandemic, against anti-Black and anti-Indigenous state violence, and the related re-emergence of fascism, *Borom Sarret* can be seen to be, in Walter Benjamin's terms, blasted out of the continuum of history and shot through with "now time" (Jetztzeit). Such "now time" crystallizes in at least three ways.

First, as I have recently argued in my book *Spectres of Fascism* (Pluto, 2020), the return of fascism provokes a reconsideration of Aimé Césaire's theory of endocolonialism—fascism as the application of techniques of domination perfected in Europe's African and Asian colonies to the European context itself. The fascist imaginary was anchored to German and Italian colonial projects in Africa and the US Republic's genocidal westward expansion.

Second, at the same time, however, the brutalities of policing cannot be reduced to "White supremacy" alone, but must be also situated in class and gender relationships. The role of the police is to protect private property, which is to say the separation between the worker and the means of production. Separation from the means of production is the condition for the possibility of exploitation as workers must sell their labour power which is rendered abstract, temporally quantifiable, and measurable. *Borom Sarret* makes this explicit insofar as the wagoner is literally deprived of his own means of production at the moment that his cart is confiscated. The abstract violence of this gesture forces his wife—both means of production and worker in one—into the nexus of the sex industry in order to engage in socially reproductive labour.

Third, the police also, of course, maintain the specifically *spatial* separation common to virtually all African cities, that between the natives' quarters or the "Medina," on the one hand, and the settlers' quarters the "Plateau," on the other, which, as Sembène shows us, is taken over by the post-colonial African bourgeoisie.

Today, in the West, but especially North America, we see the intimate ties between fascism, on the one hand, and an increasingly militarized police apparatus. Here, we see the brutal over-policing of Black people in US and Canadian inner cities and Indigenous peoples in their own territories, in particular. What Fanon calls the "well-built town" of the settler anticipates the White "gated community" fortified by increasingly privatized and militarized police forces which function, for all intents and purposes, like armies of occupation in the precincts of the poor and indigent. A society of separation; a society of the post-colonial spectacle. This becomes especially clear in Kimberly Jones's powerful analysis of looting entitled "How Can We Win?"

The Consummation of Consumption

After discussing a number of Jones's arguments, I turn to Guy Debord's reflection on the Watts Uprising of 1965 which stands in a certain relationship to the events of May 1968 and also, of course, to what happened in Minneapolis in the wake of the murder of George Floyd (Fig. 2.5).

What constitutes the axis is, I would suggest, that these are three *events* in the sense meant by Alain Badiou. For Badiou, the event signifies a moment at which the impossible becomes possible and the moments comprising this axis are three moments at which time capitalist society's own fantasy or dream about itself is profoundly disturbed.

For reasons of time, I will cut straight to the chase and identify four aspects of Jones's analysis that I find especially noteworthy:

Fig. 2.5 The consummation of consumption: stills from "How Can We Win," YouTube, June 1, 2020, still

1. Jones begins with an attack on the condemnatory response of wealthy Blacks to the uprising which is, to refer to Langston Hughes, "Go Slow." Jones is clear that she is viewing things not from the perspective of Black people per se but from the perspective of *poor* Blacks. So, her focus isn't simply on the difference between Identities, that is, Black and White, but also the differences with them, that is, the differences within the Black community, which includes substantive class differences and conflicts within this community over the very meaning of the *event*, itself. Here it is possible to argue, I think, that those middle-class Blacks who condemn the protestors, rioters, and looters, and, in the process, offer an apology for an unjust and violent social order, like colonial and post-colonial elites, *identify with the aggressor* as a response to the traumatic material of history.
2. Jones's discussion of the boardgame Monopoly as analogy for the failure of the social contract in the United States is powerful and her invocation of Tulsa and Rosewood shows the extent to which Black socio-economic and political gains have resulted in what Terry Smith calls a White backlash or Whitelash for short. Donald J Trump may be regarded as the personification of this in his rancorous attempt to systematically undo the legacy of the Obama White House, including and especially the Affordable Care Act, even if, at the end of the day, as critics like Cornel West and Keeanga-Yamahtta Taylor, among others, rightly point out, under Obama the socio-economic conditions of Black Americans actually worsened to a greater extent than their White counterparts.
3. Jones claims that the social contract is broken. Here I would challenge her claim somewhat with reference to Jamaican political theorist Charles Mills's concept of the *racial contract*. This is the idea that the contractarian tradition from Hobbes through Rawls is premised upon an unacknowledged exclusion of Black and Brown people and therefore a hidden yet no less consequential White Supremacy. One could say that this is the repressed content of political theory.

   For example, the Lockean idea that North America was *terra nullius*—that the land was "nobody's"—lent legitimacy to the settler colonial project—which, by the way, was a project that consisted of little other than *looting* on a grand scale. So perhaps it's not a matter of the contract being broken at all but functioning as it should. The point is not that the liberal-democratic social contract ought to be adhered to by way of equal treatment under the law but fundamentally rewritten to move beyond the premises of liberal democracy, itself. Its deferred dreams are dreams deferred infinitely for Black and Indigenous peoples. Langston Hughes: "The prize is unattainable."

4. The last and, in my view, most important claim worthy of note is that her rejoinder to wealthy Blacks takes the form of a defence of the figure of the "looter," which she defetishizes, by refusing a fixation on *what* it is they're doing, that is, egregiously smashing and grabbing commodities, but *why* they are doing it. And this is an indictment of US capitalism, if not capitalism as a whole. Again, as Marx indicates with his concept of primitive accumulation in Chapter 26 of Capital, this is a system that is made possible by systematic looting (embodying the real primitivism that is then projected onto its victims):

> The discovery of gold and silver in America, the extirpation, enslavement and entombment in mines of the aboriginal population, the beginning of the conquest and looting of the East Indies, the turning of Africa into a warren for the commercial hunting of black-skins, signaled the rosy dawn of the era of capitalist production.

I want to turn briefly now to Guy Debord's article entitled "The Decline and Fall of the Spectacle-Commodity Economy," from the Issue #10 of *Situationiste Internationale* published in March, 1966. This will help us to draw out some of the radical implications of Jones's analysis.

Like Jones, Debord draws attention to the almost universal condemnation of the Watts Uprising. He singles out remarks by the head of the NAACP at the time Roy Wilkins, who argued that the riot "ought to be put down with all necessary force." Like Jones, Debord understands the uprising not in racial but in class terms, referring to MLK Jr's statement in a recent Paris lecture that Watts wasn't a "race" but a "class" riot. What drives the Blacks of Watts is proletarian consciousness, according to Debord, which means consciousness that they neither are masters of their own activities nor of their own lives.

The crux of Debord's analysis aims at an inversion of the characterization of looters as the embodiment of animalistic drives (as Sharon reminded us). He does so by deploying a concept that he would elaborate in his most famous book two years later, which, in fact, gave direction to the events of May, 1968, and this is the concept of the *spectacle.* According to Debord, the spectacle is capital accumulated to the point where it becomes an image. The spectacular society is the society that creates amidst real misery and deprivation the appearance or fantasy of affluence and abundance. (It is a systematic "turning of a blind eye" as Maxine puts it.)

The spectacle represents a new level of the fetishism of the commodity form which is an object with a certain use value that satisfies determinate human needs but that is, nonetheless, produced in order to realize its exchange value or profit. For Debord, the looters, far from being animals, represented a human response to dehumanizing conditions, namely, the fact that capitalist society, characterized by generalized commodity production, is a society in which relations between things appear as relations between people and relations between people resemble relations between things.

By challenging the almost theological sanctity of the commodity, the looters re-establish human relationships grounded in gift and potlash economies. For Debord the racist and colonial "hierarchy" of the society of the spectacle, people of colour, but particularly Black people, are reduced to the status of things. Insofar as the looters directly circumvent the logic of exchange with the demand for use, which is to say, the satisfaction of needs, however false such needs may be. He argues, and I quote: "The flames of Watts *consummated* the system of consumption. … Once it is no longer bought, the commodity lies open to criticism and alteration, whatever particular form it may take." Yet, such flames immediately call into action the police. The policeman is

> the active servant of the commodity, the man in complete submission to the commodity, whose job it is to ensure that a given product of human labor remains a commodity, with the magical property of having to be paid for, instead of becoming a mere refrigerator or rifle—a passive, inanimate object, subject to anyone who comes along to make use of it. In rejecting

> the humiliation of being subject to police, the blacks are at the same time rejecting the humiliation of being subject to commodities.

The social contract, to reiterate, is not broken but functions all-too well for it is a contract geared to the maintenance of private property.

Returning to the question I started with, namely: the possibility of cross-racial solidarity, it is of vital importance to grasp the particular and universal significance of the uprisings and in the process to make of it more than a "racial" event, for this is exactly what the far-right want. Rather, we must situate the uprisings that we're seeing within the larger context of a society in which inequalities are deepening; it is also important to place recent developments within in the context of a history of social struggles, from Watts in 1965 to Paris in 1968 to Minneapolis in 2020. It is vitally important to understand extreme forms of police violence not as effects of a mystical, transhistorical White supremacy, but rather as a manifestation of a racism that flows from the vicissitudes of a social order mediated by the commodity-spectacle, grounded in the sanctification of private property under deepening forms of socio-economic inequality that nonetheless hits Black and Indigenous communities especially hard.

This social order is a historical one—an order that came into being and one from which it is possible for us to emancipate ourselves.

Conclusion

Contemporary fascism emerges from the phenomenon of accelerated global migration flows resulting from the economic, social, and political violence (new forms of primitive accumulation) attendant upon the global reconstitution of the relations of production. It responds to the increasing ontological insecurity of citizens of these states—inestimably bolstered now by the pandemic—whose fear is increasingly and effectively mobilized against myriad *strangers* turned into *enemies*. Such mobilization is based on the acute awareness that, under the late form of neoliberalism, the line between citizen and migrant, parvenu and pariah, in other words, "genuine" and "superfluous" humanity, is increasingly blurred. Capitalism has always embodied a sacrificial logic and this lies at the heart of its authoritarian potential today. This logic deepens when workers, particularly White workers, hand in hand with the lower middle class, come to identify with rather than contest the power of the aggressor.

Yet, as dire as the situation may be, there are hopeful signs of growing labour militancy, as was recently demonstrated by striking workers at Amazon, Instacart, Shipt, and Whole Foods on May Day, who protested what they considered to be their employers' woefully inadequate responses to the pandemic. The global health emergency, moreover, has demonstrated that the integrity of societies cannot be indexed to the prosperity and well-being of its most affluent but most indigent members. It has decisively shown that healthcare cannot be tied to conditions of employment but must be understood, as Bernie Sanders repeated over and over again in his bid for the Democratic presidential nomination, as a basic human right. It has highlighted the nihilistic illusions of the "possessive individualism" on which shifting sands of the entire neoliberal order are based. It has seriously revived, with great urgency, the discussion of the admittedly fraught and contested idea of Universal Basic Income. The pandemic has doubtlessly, as I have argued, constituted an opening for a further authoritarian consolidation of power but, at the same time, it has also opened space for imagining a very different kind of society. Which path we take will be a matter of organizing, which is to say, *political engagement and struggle*.

References

Adorno, Theodor W. 1998. Critique. In *Critical Models: Interventions and Catchwords*, trans. Henry Pickford, 181–88. New York: Columbia University Press.

Amin, Samir. 2014. The Return of Fascism Within Contemporary Capitalism. *Monthly Review*, 1 September 2014. monthlyreview.org/2014/09/01/the-return-of-fascism-in-contemporary-capitalism/.

Badiou, Alain. 2015. *The Communist Hypothesis*. London: Verso.

Beckett, Samuel. 1964. *Endgame: A Play in One Act Followed by Act without Words, A Mime for One Player*. London: Faber and Faber.

Cavell, Stanley. 1969. Ending the Waiting Game: A Reading of Beckett's *Endgamei*. In *Must We Mean What We Say: A Book of Essays*, 115–162. Cambridge, UK: Cambridge University Press.

Césaire, Áime. 1972. *Discourse on Colonialism*. Trans. Joan Pinkham. New York: Monthly Review Press.

Debord, Guy. 1966. The Rise and Fall of the Spectacle-Commodity. https://www.cddc.vt.edu/sionline/si/decline.html.

Financial Times. 2019. Vladimir Putin Says That Liberalism Has Become Obsolete. June 27. https://www.ft.com/content/670039ec-98f3-11e9-9573-ee5cbb98ed36.

Finchelstein, Federico, and Jason Stanley. 2020. The Fascist Politics of the Pandemic. *Project Syndicate*. https://www.project-syndicate.org/commentary/coronavirus-fuels-fascist-politics-by-federico-finchelstein-and-jason-stanley-2020-05.

Foucault, Michel. 2010. *The Birth of Biopolitics: Lectures at the Collège de France 1978–79*. London: Picador.

Gandesha, Samir. 2018. Identifying with the Aggressor: From the 'Authoritarian' to the 'Neo-Liberal' Personality. *Constellations* 25: 147–164.

———. 2020. Introduction. In *Spectres of Fascism: Historical, Theoretical and Contemporary Perspectives*. London: Pluto.

———. 2020a. Canada's Battle against First Nations Shows Slide into Authoritarianism. https://truthout.org/articles/canadas-battle-against-first-nations-shows-slide-toward-authoritarianism/.

———. 2020b. Ousman Sembene and the Part that Has No Part. *openDemocracy*. https://www.opendemocracy.net/en/ousmane-sembène-and-the-part-that-has-no-part/.

———. 2020c. The Consummation of Consumption Parts 1 and 2. *openDemocracy*. https://www.opendemocracy.net/en/consummation-consumption-1/; https://www.opendemocracy.net/en/can-europe-make-it/consummation-consumption-2/.

Gardner, Amy, Kate Rabinowitz, and Harry Stevens 2021. How GOP-Backed Voting Measures Could Create Hurdles for Tens of Millions of Voters. *Washington Post*, March 11, 2021. https://www.washingtonpost.com/politics/interactive/2021/voting-restrictions-republicans-states/.

Gidla, Sujatha. 2020. We Are Not Essential, We Are Sacrificial. *New York Times*, 6 May. Accessed January 9, 2020. https://www.nytimes.com/2020/05/05/opinion/coronavirus-nyc-subway.html.

Hayek, Freidrich A. 2007. *The Road to Serfdom*. Chicago: Chicago University Press.

Heidegger, Martin. 1962. *Being and Time*. Trans. John Macquarrie and Edward Robinson. San Francisco: Harper and Row.
Jones, Kimberley. 2020. How Can We Win? *YouTube*, June 9. Accessed January 9, 2020. https://www.youtube.com/watch?v=llci8MVh8J4.
Mbembe, Achille. 2017. *Critique of Black Reason*. Durham, NC: Duke University Press.
Mills, Charles. 1999. *The Racial Contract*. Ithaca, NY: Cornell University Press.
Piketty, Thomas. 2020. *Capital and Ideology*. Trans. Albert Goldhammer. Cambridge, MA: Belknap Press.
Rancière, Jacques. 2013. *The Politics of Aesthetics*. London: Bloomsbury.
Riley, Dylan. 2018. Editorial: What Is Trump? *New Left Review*, No. 114, November–December: 5–31.
Rousseau. 1968. *Social Contract*. London: Penguin.
Žižek, Slavoj. n.d. The Palestinian Question: The Couple Symptom/Fetish Islamo-Fascism, Christo-Fascism, Zionism—Mieux Vaut Un D.sastre Qu'un D.s.tre. lacan.com. Accessed August 16, 2019. www.lacan.com/essays/?page_id=261.

3

Technology and Biopolitics: A Deleuzian Perspective

Sin Heng Tony See

Agamben on COVID-19 and the Rise of Biopolitics

In a recent interview Giorgio Agamben pointed out that while the infection rate for the coronavirus may be high the actual death rate from the virus is small when compared to the other diseases (Agamben 2020). He highlighted the fact that its fatality is actually less than children's death rate from annual influenza, "In 10–15% of cases a pneumonia may develop, but one with a benign outcome in the large majority of cases. It has been estimated that only 4% of patients require intensive therapy." When we look at the scientific evidence we realize that while many have been infected, a significant number of those who actually succumbed to the disease did so because of prior medical conditions. It is always possible to survive the virus provided there is adequate and timely medical care. Agamben states "If this is the real situation, why do the media and the authorities do their utmost to spread a state of panic, thus provoking

S. H. Tony See (✉)
National University of Singapore, Singapore, Singapore

S. S. Das, A. R. Pratihar (eds.), *Technology, Urban Space and the Networked Community*, https://doi.org/10.1007/978-3-030-88809-1_3

in the authentic state of exception with serious limitations on movement and a suspension of daily life in entire regions?" Agamben suggests that the panic induced by the media can bring about an authentic "militarization" of social life. This is confirmed as various countries imposed movement controls, quarantines and surveillance on the general public. Whereas external enemies once served as a legitimation for tighter social control within domestic sovereign territories, now "health" becomes a factor in social control. Such a concept as "health," however, is vague as it enables the state apparatus to rapidly extend its police powers, movement controls and surveillance to all regions within its territorial boundaries. In short, what Agamben warns of is the potentiality of the COVID-19 pandemic as an occasion for the state to expand the "state of exception" to new areas (Agamben 2020).

Agamben's warning about the rise of biopolitics against the background of the COVID-19 global pandemic follows the overall trajectory of his thought on sovereign "bare life" and the "state of exception." In *Homo Sacer: Sovereign Power and Bare Life* (1998), he has articulated the idea that the state has always gained control over human lives by using juridical means to remove them from their "political life" (*bios*) in order to plunge them into a state of "bare life" (*zoe*). When people are stripped of their political life and reduced into "bare life," they are no longer considered human beings, much less as citizens of the *polis* that are able to enjoy certain protections and privileges from the state. Nor can these people be sacrificed to the gods because they are not technically speaking "human beings." They essentially find themselves located within a zone of indistinction in which neither human nor divine laws apply, so that they can now be killed by the state without violating any law (Agamben 1998, 81).

This "state of exception" is not restricted to right-wing political organizations but can also be found in the democratic-revolutionary traditions. Agamben notes that when Napoleon introduced the *state of siege* in his decree of December 24, 1811, during the Revolutionary period, he essentially introduced the state of exception into the city because now the emperor could declare whether or not the city was in a "state of war" (Agamben 2005, 5). This essentially gives the emperor absolute power over the life and death of his subjects. While the state of exception was originally conceived as a response to exceptional circumstances that endangers the population from without, subsequently developments saw this doctrine being used to defend the *status quo* from "enemies" from within the population itself. Secondly, this concept was also expanded from wartime situations to peaceful conditions, so that extraordinary police measures are now authorized *at all times* in the city to cope with "internal sedition." In such "state of exception," the executive obtained "full powers" (*pleinspouvoirs*) in issuing decrees that have the force of law.

This state of exception and its biopolitical programme continue to operate in modern times throughout the two world wars and its aftermath. In fact, it continues to operate in contemporary twentieth-century politics as the often practiced but seldom acknowledged paradigm for modern governments. Agamben highlights in his short text *State of Exception* (2005) that when the president of the United States issued the "military order" in 2001, he in effect authorized the "indefinite detention" of noncitizens who are suspected of involvement in terrorist activities. This is to be brought before a trial by "military commissions" that are formed in the United States. This power to try foreign citizens before an American military court expands the power of the attorney general, who was already invested with the power to "take into custody" any alien suspected of activities that endangered "the national security of the United States" by the USA Patriot Act issued on October 26, 2001. What was unique with this "military order" is that it also erases any legal status the individual may have, producing a legally unnameable and unclassifiable being who in the language of classical political theory, can be killed without either infringing human or sacrificial laws (Agamben 2005, 3).

Agamben maintains that when the exception becomes the norm, what is undermined is none other than parliamentary democracy itself. This is because laws of this nature, which belong to exceptional circumstances of necessity or emergency, conflict with the fundamental principles, hierarchy and regulation of democratic institutions. As legal thinkers may say, "when what was merely a temporary power becomes a systematic and regular exercise, it can only lead to the 'liquidation' of democracy" (Agamben 2005, 6–7).

Agamben's claim that biopolitical power has been expanding throughout human history and that we need to be mindful of how it may continue to expand its powers during this COVID-19 pandemic needs to be taken seriously. This is especially so when various governments throughout the world channelled their resources to the development and deployment of surveillance technologies instead of exploring other options as a means of containing the virus. In the early months of 2020 following the WHO announcements that we have a global pandemic at hand, there is not simply an obsession with lockdowns and movement control of the populace but also the expansion in the use of tracking devices, artificial intelligence and biometric facial recognition technologies and so on. The fact that the state has decided to go the way of surveillance technologies without seriously considering other options for containing this pandemic needs to be considered carefully by those who deny the dominance of the biopolitical dogmatic image-of-thought. A second point to note in this COVID-19 pandemic is the introduction of "health" as a new function in national security. While security has been a concept that was traditionally associated with military and economic discourse, what we are seeing is that health is increasingly militarized as a concept justifying greater police powers and control. That is, health in the post-COVID-19 world will be weaponized into a national security issue—that is, justifying police power and surveillance. Aren't there reports of how we are to be injected with a microchip that will track our movements and even our heart rates? It is interesting how in the pandemic situation, even up to March 2020 way before the discovery of the vaccines capable to preventing and even curing the virus, the structures of surveillance and social control are almost everywhere in place. It would suggest that the state is more obsessed with social control than with social safety from the virus. In short, it is inevitable and quite clear that COVID-19 provided a new occasion for the rise of biopolitics.

Deleuze and Societies of Control

The current situation, one characterized by lockdowns, movement controls and the rise of tracking devices and biometric surveillance, seems to be a caricatured performance from the page of one of Agamben's books on biopolitics. His thesis about the rise of biopolitics, whether he got his numbers correct, seems to be correct when various countries rushed to adopt movement controls and surveillance technologies in order to contain the spread of the virus. In this section, we take into consideration the technological regime that affords the rise of such a politics. Deleuze's theory of "societies of control" seems to provide the theoretical grounding for the rise of such a regime. Deleuze's focus was not on biopolitics, nor could he have foreseen the rise of any pandemic that approximates the scale of the COVID-19 global pandemic. The COVID-19 pandemic is in a sense something unthinkable in human history, something that is barely remembered in biblical stories nowadays. However, Deleuze's theory of "control societies" does give us some hints as to what type of technological regime this biopolitical future might look like. In his "Postscript to Societies of Control" (1992) he intimates that we are already entering the stage of "control society." This "society of control" is specifically described by Deleuze to be distinct from Foucault's "disciplinary society." In Foucault's "disciplinary society" institutions are arranged in such a way that throughout their lives, members of the society have to pass through physical boundaries and enclosures, and as each of these enclosures has its own rules, there is a clear line to cross in order to leave behind an enclosure and to enter another enclosure. What is obvious in this system is that every time a person enters a new environment, they will have to embrace a new logic, and what they have experienced and embraced previously will potentially become irrelevant in the context of their new environment. Some examples of these spaces and closed environments are the main institutions of the family, school, barracks, factory, hospital and prison.

Foucault's idea of disciplinary society is derived from the idea of "panopticon" in his earlier writings. The word "panopticon" stems from the Greek word "pan" which means "all" and "opticon" which means "visual." The idea behind the "panopticon" is the idea of a physical space, a semi-circular physical building to be constructed in such a way that it allows prisoners to be easily watched from a central tower. The circular design of the building includes a tower in the centre in which a guard is able to keep all the prisoners in full view, and at all times. As such, according to Foucault, "in the peripheric ring, one is totally seen, without ever seeing; in the central tower, one sees everything without ever being seen" (Foucault 1977). One result of this design is not merely physical control but also the psychological control exercised on the prisoners. The possibility of being constantly watched and being under surveillance due to the nature of the panoptical prison causes prisoners to *discipline* themselves and regulate their own behaviour. In other words, there is a reproduction of a certain form of subjectivity when the prisoners spend their time in the prisons. Foucault's idea of disciplinary society may be extrapolated to a wider picture, to refer to how an entire society, and the subjects found within such a society, may be structured using such a logic. In his lecture at the College De France from 1978 to 1979, this comes to be known as "biopolitics" and it essentially refers to a form of governance based on governing "populations" instead of peoples (Foucault 2008).

For Deleuze, we are already moving away from a society based on the logic of "discipline" to a society governed by the logic of "control"—a "control society." In a control society there is no longer a clear physical boundary between spaces that could be crossed, nor is there a central "site of control." Deleuze states that while the concept of "discipline" as a form of power relies on physical boundaries, the society of "control" blurs the lines of these boundaries and relies on a different sort of control mechanism which may be called "a modulation." Deleuze's idea of modulation is defined as a "regulation according to measure or proportion" (Modulation [Def. 2], n.d.). Unlike the static, unchanging nature of institutions in disciplinary societies, a "modulation" is control according to a measure or proportion. Its control is fluid and flow freely through what previously was separated, closed spaces. This means that unlike in a disciplinary society, a person can no longer "end" or "restart" at any point of his life. Instead, he will not be able to say he is done with anything. This is because there is no finishing line since every stage is now interconnected. Deleuze gives the example of education. In disciplinary societies, one can say that he is done with the "school" stage of his life; he may complete his three years undergraduate programme and graduate with a degree and so on. However, in societies of control where education in schools has morphed into lifelong learning, and now "online learning" and "distance learning," a person will essentially never be done with "school."

Deleuze states that "societies of control" will also work not through institutions like factories and schools but instead through "continuous connections" as well as "instant communication" which can be seen as the key apparatuses of a control society (Deleuze 1995). As such, one can infer that Deleuze was making references to internet, communication and surveillance technologies when he spoke about the shift in modality of power where incessant communication, information processing as well as the fluidity of power and control go way past institutional boundaries, leading to the formation of a society of control (Deleuze 1995). What is important now is a "code"—he theorizes that the "numerical language of control is made of codes that mark access to information or reject it" (Deleuze 1992, 5). Codes have become the basic units in control societies. This is in radical contrast to the concept of the individual in disciplinary societies where the individual is indivisible, the smallest unit of society. In control societies, individuals are now *dividuals*, who can be further divided or broken down into code which is processed by computers. Deleuze identifies computers as the machines of control societies, a technological evolution that is a "mutation of capitalism" (Deleuze 1992, 6). He follows by giving a brief introduction to the form of capitalism that prevails in control societies, an entirely different animal from the iteration that was prevalent in disciplinary societies. In societies of control, capitalism deals with services and stocks rather than goods and a single owner.

Unfortunately, while Deleuze's theory of societies of control illuminates a form of society that is to come with the advent of new technologies, his thoughts on societies of control are also severely limited to a short abstract such as "Postscript on Societies of Control" and brief descriptions in another writing called "Control and Becoming" in his interview now published as *Negotiations* (Deleuze 1995, 169–176). This makes it extremely difficult to carve out with some degree of certainty Deleuze's view with regards to this new form of organization, and scholars like Hardt have long tried to articulate the concept based on the limited information available (Hardt 1998). In Foucault's disciplinary societies, each and every institution persists through its own logics and rules and within these institutional walls, an individual can be seen to be partly protected from the rules and powers of other institutions (e.g., a teacher cannot discipline or have power over a student that has gone home) (Hardt 1998). However, the shift towards a society of control from a disciplinarian society has seen these institutional walls being broken down—an individual is always at the workplace, always in school, always in prison and so on (Hardt 1998). The dissolution of institutional walls means that the person is no longer trapped within the enclosure, and yet as Deleuze suggests, this freedom and openness can also become a form of entrapment. Previously, when one was trapped within a physical enclosure, one can also become free as long as one leaves behind this enclosure. Now that the enclosure does not exist, it seems as though one would be completely free. But the open spaces, according to Deleuze, only give the illusion of freedom. Rather than trapping things within a confined space, control works via actively capturing people and objects. He warns that control has become more insidious, and that it is up to youths to realize that despite the freedom they might feel, they are still within the now-invisible grip of control.

While Deleuze may have suggested that societies of control will replace the societies of discipline, there is no indication that these two cannot overlap and exist together in the same space. In the essay "Postscript" he also brings up the idea that in the societies of control, while it may render some societies where the binary logic of disciplinary societies becomes obsolete, the logic of societies of control may also operate in a way that this disciplinary logic is not rendered useless or ineffective but instead becomes more fluid across society. It is illuminating, however, to note that Deleuze concludes his description of societies of control as a "progressive and dispersed installation of a new system of domination" (Deleuze 1992, 7).

Surveillance Technologies and the Biopolitics of Control

Agamben's idea of biopolitics as the governance of "bare life" and population and Deleuze's idea of distributed power in "control societies" bear little resemblance to each other. One is based on juridico-legal writings found in the western legal-political tradition while the other is based on rhizomatic and immanent reflections in the western philosophical tradition. Nor are we to assume that they are addressed to each other. Agamben was describing the rise of biopolitical thinking based on juridico-legal texts in the western political tradition while Deleuze was interested in envisioning a new borderless world in contradistinction to Foucault's idea of a "disciplinary" society. Nevertheless, there is a remarkable resonance in their writings, one which becomes clearer when we consider developments made in surveillance technologies in the name of biopolitical control. The rise of the new biopolitical control must be seen on the basis of surveillance technologies, and this in turn must be seen in the context of decades long developments in new internet, communication and intelligence technologies. These seemingly innocent and naïve developments in technologies were seen as liberating individuals in society, not as instruments of surveillance and enslavement of workers. In fact, the new technologies were received with much enthusiasm by scholars from various fields, many who can be called "digital optimists" who saw the rise of the new technologies in terms of the long awaited dawn of a new global transformation process that will bring democracy and equality to people across the globe. They envisioned that the new technologies would bring about a series of political, social and economic transformations that will bring about the end of authoritarian governments and oppressive inequalities that are prevalent in capitalist economies. In fact, the new technologies are also seen as harbingers of a new form of "network society." Here, we will refer to three thinkers who influenced the reception of the new technologies: Marshall McLuhan (1962), Howard Rheingold (1993) and Manuel Castells (2000).

McLuhan (1962) is usually known in media and communication studies for his idea that "the media is the massage." However, what we need to take note is that he is also known for being a strong supporter of the thesis that new technologies would bring about the emancipation of individuals. This is made possible when new media and information technologies allow them to gain instant access to huge amounts of information in an instant, to communicate effectively and form new relationships with others across vast distances and eventually to form a new "Global Village." The idea was that the new communication technologies while having issues of "massaging" or influencing the perceptions of its subjects are also able to create a complex web of electronic communications system, in a similar fashion to the human nervous system, that integrates the entire world, allowing people from different countries and nations to connect and share information with others across vast distances much faster and more intimately, bypassing physical boundaries and time differences, practically "shrinking" the world. Thus, McLuhan essentially proposed that we will enter a new world where global resources are shared, and economies and nations are interdependent (McLuhan 1962).

This optimism with regards to the new technologies was shared by other media and technology scholars, one of the more prominent being Howard Rheingold (1993). Rheingold proposed that we will enter a future filled with infinite possibilities for human development through a global network. He suggested that with the development of new technologies we can create virtual communities with like-minded individuals for any reasons, ranging from social purposes to discussion of politics, in other worlds, we are in an age of what he calls "Virtual Communities" (Rheingold 1993). These *Virtual* Communities are social aggregations of people connected through computer networks, where they establish public discussions, and with sufficient investment of human feelings, form personal relationships. Rheingold proposed the possibility that these citizens-designed and citizens-controlled network would eventually give birth to a new form of online democracy, or what he called "an electronic algora," that decentralizes political communication and revitalizes citizen-based democracy (Rheingold 1993).

In addition to Rheingold, Manual Castells, in particular, was famous for articulating the outlines of a new "Network Society" in which people throughout the world are connected through the development of new forms of network and communication channels (Castells 2000). Castells introduced the idea of real virtuality in his major three-volume work *The Information Age*: "It is a system in which reality (that is, people's material/symbolic) existence is entirely captured, fully immersed in a virtual image setting, in the world of make believe, in which appearances are not just on the screen through which experience is communicated, but they become the experience." This level of experience is not separate from reality but *identical* with it (Castells 1996, 373). In his later work *The Internet*

Galaxy (2000) Castells restated his stance to claim that a virtual community is a form of social reality. Virtual communities do not *replace* but change social relations. It has a positive impact on social relations because it enhances democratic possibilities and offers people a more communicative means of ordering relations, that is, the use of email by higher educated helps to sustain networks of friends and family. What we have is actually a "thin" community that adds a dimension to people's lives. He states "Networks are built by the choices and strategies of social actors, be it individuals, families, or social groups. Thus, the major transformation of sociability in complex societies took place with the substitution of networks for spatial communities as major forms of sociability" (Castells 2000).

While these scholars articulated a vision of a beautiful shared communal and democratic future based on new technologies, other scholars argued that the new technologies would bring about a more equitable society. These scholars, sometimes known as "digital utopians," held the view that the new technologies in internet, communication and surveillance have the ability to bring about a *post-scarcity economy*. This is an economy where people, supported by new technological developments, are able to work less and yet be more productive. The system once called "Taylorism" essentially proposed the use of scientific methods to improve the productivity and efficiency of the workers. Furthermore, the introduction of new computation algorithms and new "quantum" technologies allowed for considerable computations to be done in an instant, while the invention of global networked communication allows for messages to be transferred quickly over distances. These scholars believe that eventually these new information technologies when combined with new robotics and AI would make the more labour-intensive types of work a thing of the past, there would be higher wages and a rise in the overall quality of life. These technological and digital utopians held an implicit faith in *trickle-down economics* where the overall benefits of the new technologies will be well distributed fairly in society, the immense wealth disparity between the poor and the rich in society will be closed and the expansion of this throughout the world will bring about a transformation at the global level.

While these technological optimists were entrenched in their celebration of the rise of the new technologies and the new spaces that they afforded, there are also those who argued that there is a darker side to the new technologies and the economy that it entails. They argued that while it may produce wealth more efficiently, the bulk of this wealth goes to the wealthy class. The new technologies and their new management systems seem equally capable of bringing about poverty. "Taylorism" indeed uses scientific methods to improve the productivity and efficiency of the workers, but this is usually done through the simplification of jobs and the delegation of the skills and knowledge of the workers in relation to newly developed machines and even robots. One consequence of this simplification is that it provides the economy with a ready supply of workers whether human or robotic, while it is able to produce greater efficiencies and productivity, it has also rendered each worker replaceable like a cog in a gigantic capital machine and overall more insignificant in the system. Workers in the economy have essentially become what Heidegger would call a "standing reserve." Towards the end of the nineties, a new form of management was introduced known as "Digital Taylorism" (Schumpeter 2015). This new management system saw the rise of a new form of technology that is developed in order to engage in the surveillance of workers and their movements. For instance, companies strapped an armband onto their warehouse workers' arms to measure their productivity and track their movements. Gradually, these surveillance technologies were extended beyond the boundaries of the workplace and work time and into the workers' homes and private lives. Some companies, especially social media companies, have been found to assess information of employees and potential employees that are found in their private social media as a basis for employment or advancement considerations. Secondly, the new form of *Digital Taylorism* is also exemplified through the emergence of a type of workplace in which managers and employees are expected to stay constantly connected for work-related matters in a way that goes way beyond their working hours and workplace. As a result, the personal time of individuals is shortened or eliminated given that business-related messages or emails may still be directed to them even after working hours. This contributes to the trend of longer working hours and a new social, or rather, anti-social phenomenon of people working round the clock to meet the demands of their employers. In other words, the irony behind the new technological era is clear, the technology that is purported to liberate individuals by allowing them unconstrained connectivity and access to information has in turn come to enslave them (Schumpeter 2015).

The emergence of the new technologies does not only mean increasing the gap between the rich and the poor, but also the dissolution of the middle class in society. This is largely due to the displacement of workers by new robots and through the process of outsourcing work overseas to developing economies. Manufacturing companies are known to have replaced thousands of workers with robots. This is often legitimized by the rise in labour cost, but it is usually due to the more banal instrumental reason that robots are able to work longer hours and with more standardization to ensure a higher rate of capital accumulation. Furthermore, the new surveillance technologies also allow big multinational corporations to coordinate their operations across multiple geographies. This results in an increase in market and size, but more importantly, it also allows the company to outsource their jobs to countries that have lower wages levels, and to countries with little or no labour unions and regulations. Less developed countries are known for having less regulations and a lack of union protection, and this enabled companies to set up "sweatshops" in third world developing countries where the political organizations are draconian, where working conditions are oppressive. In addition, companies are increasingly employing contract staff, which again allow them to bypass unions. Unlike full-time employees, contract staff are not protected by unions and do not enjoy many benefits, which include paid vacation, medical and insurance. Indeed, companies now are able to dramatically reduce their costs by depressing wages and employee-related expenses due to the shifting power balances afforded to them by the advances in the new technologies.

Stiegler and the New Proletarianization of Consumers

Bernard Stiegler (2014) has suggested that in addition to the rise of Digital Taylorism, there is also a proletarianization of consumers. In *Symbolic Misery—Vol. 1: The Hyperindustrial Age* (2014), Stiegler uses the idea of "symbolic misery" to describe the hyperindustrial society as a society which corrupts consciousness by ingraining in people a standardized idea of desirability. There is one prescribed standard of desirability that is propagated endlessly via the media and various dominant institutions. Diversity in the truest sense of the term is a thing of the past, a romantic imaginary whose reminder can only cause further misery. The idea of a free-thinking citizen able to make critical decisions for himself or herself about the social and political organizations that he or she needs in effect becomes a thing of the past. Through the extensive reach of mass media and new digital technologies, corporations now have the power to exert considerable influence on consumers, slowly structuring the way they think, perceive and assign meanings to the things around them. This can be seen in how some technological companies fiercely tout their products as lifestyle choices and strengthening that message through a range of communication channels that are made possible by the advancement in surveillance and propaganda technologies.

Stiegler (2014) argues that due to this proletarianization process in the hyperindustrial age, the consumers still experience some sort of "freedom," but this is not the freedom that Heidegger or Deleuze would talk about. Instead, what they experience as their "freedom" is nothing more than the feeling of being "discharged" from the duties of thinking and shaping their own life, and they are as such reduced to mere numbers on *a* diagram representing consumer demand and buying power. Faced with the constant bombardment of marketing materials, consumers are slowly discouraged from thinking and encouraged to desire for the never-ending list of products available in the market. According to Stiegler, this process robs modern consumers of their *savoire vivre (knowledge of living)* and as a result, destroying their *joie de vivre* (*joy of life*). As such, it would suggest that the development of the new information technologies destroys individualization and the ability of consumers to make informed decisions. Following that, the consumers' perception of choice may in fact be illusory due to media conditioning (Stiegler 2014).

Although the internet is able to democratize authoritarian regimes by empowering the masses, we need to see that the same technologies are also used as a means to engage in mass surveillance, political repression and the spreading of propaganda. The United States, for example, built a \$1.7 billion facility at Utah to store data on the Americans' use of the Internet (Carroll 2013). In Britain, similarly, Theresa May, Home Secretary of the United Kingdom, proposed harsh surveillance and spying laws, requiring Internet companies to submit communications data of their users to the government (Stone 2015). Indeed, given the tendency for the government to have a direct or indirect influence on the type of information to be disseminated, it is unclear whether the development of the new technologies actually does allow people access to more information, or rather more accurate and unbiased information. In addition, the omnipresence of an authority figure surveying the cyber landscape raises doubts as to whether ICT truly emancipates people (Stone 2015).

This current belief that the new technologies would enable the free flow of information and democratize societies, therefore, seems to be quite naïve because it does not consider who is in control of this information. History already teaches that the invention of the printing machine in the fifteenth century by Gutenberg enabled the mass production of texts and messages and this facilitated a more efficient manner of disseminating information and knowledge among populations, which in turn brought about the demise of the previous feudal class arrangement. However, the information that was "freely" disseminated through the printing press came to be controlled by the ruling elites, by those with enough financial and political clout to control the transmission of information. Essentially, information and printing came to be dominated by the dominant class in society, and this gave rise to the dissemination of information and ideas that were favourable to that class, and eventually the commodification of the press. Fast forward a few hundred years, and it is quite apparent that the information disseminated in society continues to be dominated by those with the financial resources. Information is not entirely "free" but controlled by the ruling elites. A case could be made for the thesis of "freedom of speech" and how it brings about scientific progress and general happiness in society. However, these are prescriptive statements without any descriptive force in what is occurring today. Already by the 1980s, 50 different corporations in the United States control 90% of the mass media. By 2011, that number became six and in 2015 it was five: Time Warner, Disney, Murdoch, Bertelsmann and Viacom, all large conglomerates by any measure.

Unlike Foucault's disciplinary society whereby power and control are held over a person's mind and physical body, a society of control is more focused on a human being as a "data set" or "data doubles" (Lyon 2001). In Deleuze's words, "what is important is no longer either a signature or a number, but a code: the code is a password" (Deleuze 1995). The signature and the number he is referring to are human bodies while the codes have replaced the bodies and are also seen as passwords for gaining access to social locations. The control is established by no longer forcing an individual body to move to specific spaces, but via enacting specific requirements in the environment through which the body moves.

Hence, it can be said that societies of control depict the change from the disciplinary society to a society where surveillance is of paramount importance. As mentioned previously, Foucault's idea of surveillance was one that was done via the jail guard who had his eye on all the prison inmates who were fearful of his gaze and thus behaved a certain way in fear of disciplinary actions. However, in a society of control, biopolitical surveillance of the population, gathered through tracking devices and apps and consolidated in Big Data algorithm, is the new arrangement. The surveillance apparatus is as described by Foucault is now seen as outdated, to be replaced by a new form of control which is only concerned with the collection of data information about the population.

Technology and the End of Biopolitics

I believe that Delezue had a highly relational concept of technology. By this I mean that from a Deleuzian perspective, the functions of technology are relative and contextual. While it is important to acknowledge the dangers of the rise of biopolitics via various forms of technologies during this COVID-19 pandemic, we need to learn to see that these technologies are not necessarily fascist in themselves but can have liberating functions. While history has shown that over time they may to be controlled by the elites in society, there is no reason why they cannot be used to resist fascism right now. A Deleuzian perspective towards technology would have to be a creative one and not one that is restricted to a negative critique of technology. Furthermore, even as we are stunned momentarily by the impact of the COVID-19 pandemic, there is no reason why we cannot see the pandemic as a moment in history in which people throughout the world are frozen in their tracks, question what is really important, what is not, and become committed towards global and planetary causes. The pandemic, if anything, is *an Event* that opens up the world that shows that we are all interconnected and interdependent in more ways than one.

Now a critical fact about any new technologies is that they are highly ambiguous. Don Ihde, a philosopher of technology, once called this "multistable possibilities" (Ihde 2002). To say that technologies are "multistable" is not to say that they are unstable, rather, the point is that technologies cannot be reduced to their "designed functions." A designer of a specific technology may have a specific function in mind for that technology, but the actual technology may have other functions unthought of by the designer. To think otherwise is to insist that technologies can only have the specific functions that the creators created it which is ridiculous. This is the key reason why while philosophers should be wary of the rise of biopolitics via surveillance technologies, we should also be aware of the potentialities that these technologies afford us. While it is true that the new technologies have brought movement control and the loss of privacy, it is also important to note that these same technologies can also be used to resist biopolitical control. In other words, if we imagine hard enough is it possible to use the existing surveillance technologies to resist biopolitics as a politics which treats human life as "bare life" and to recreate a new planetary *Zoepolitics* which regards human life as sacred?

We may put this question in a different way: Is it possible to use the new surveillance technologies, which are also information technologies, to build collective practices that protect the health of populations without a parallel expansion of fascist forms of power? Can surveillance technologies be used to check hitherto unchecked powers? Our critique of the rise of biopolitics in an age of a possibly incessant COVID-19 global pandemic although may be well-intentioned, runs the risk missing the essential component of existing unjust biopolitical systems. It is a truism that no fascist regimes can arise without some emotional commitments given for its rise. Given this fact, we need to ask how fascist regimes and biopolitical programmes have gained ground, how did these regimes come to attract emotional commitments from both the fascists and their willing victims that allow for the effectiveness of coercion and power? We need to ask whether the surveillance technologies that have risen since the COVID-19 pandemic have opened up new possibilities by redirecting our emotional commitments to bring about positive social, political and planetary change. In other words, *is it possible to transform the present biopolitics as a politics of "bare life" into a new politics of immanence or a Zoepolitic.*

One way of thinking about this would be to think of how surveillance technologies have enabled us to become more aware of global and planetary happenings. As we are writing this for publication in the January of 2021 there are already two million deaths worldwide that are associated with the coronavirus. The numbers of infection continue to rise in each country numbering in the thousands. It was reported that there are 7000 infections each day in Japan, and many more in other countries. These "surveillance" technologies can make us rethink our priorities in life and be concerned and connected with people in other countries. They can make us more committed to the idea that the decisions for the reduction of the virus cannot be left to inefficient governments. In some cases these "surveillance" technologies can help us get connected globally and even on a planetary level, so that we organize new movements and make democratically discussed collective decisions that stand above narrow nationalistic confines. The same technologies that have been used to build a biopolitics of control could also afford us with the resources to build a new politics of human life, a Zoepolitics. The increased access to knowledge, along with the need for popularization campaigns, makes possible collective decision-making processes that are based on knowledge and understanding and not just the authority of experts, *and now thanks to the*

COVID-19 pandemic we finally come to the realization that these decision-making processes have to be planetary if we are to survive.

Conclusion

In this chapter, we have explored how the current COVID-19 global pandemic has raised fears among philosophers regarding the rise of biopolitics via surveillance technologies. The rise of such a political paradigm is doubtlessly imminent and we need to thank philosophers such as Agamben for warning us about its arrival. However, we also need to note that the rise of this biopolitical future hinges on the use of various types of information and communication technologies, with surveillance technologies being one of them. Furthermore, since the nature of technology is such that they are ambiguous and may have more than one function, it is always possible to think of alternative uses for these technologies. In other words, it is always possible, and now in fact necessary, to think of surveillance in new ways. This paper argues that while it is important to be wary of the rise of biopolitics during the COVID-19 pandemic, we should also acknowledge that technology has many potentialities and that we have to explore its potential as a basis for the rise of a new politics.

References

Agamben, Giorgio. 1998. *Homo Sacer: Sovereign Power and Bare Life*. Translated by Daniel Heller-Roazen. Stanford, CA: Stanford University Press.

———. 2005. *State of Exception*. Translated by Kevin Attell. Chicago and London: University of Chicago Press.

———. 2020. *The Invention of an Epidemic*. Published in Italian on *Quodlibet*, https://www.quodlibet.it/giorgio-agamben-l-invenzione-di-un-epidemia.

Carroll, Rory. 2013. Welcome to Utah, the NSA's Desert Home for Eavesdropping on America. *The Guardian*, June 14. http://www.theguardian.com/world/2013/jun/14/nsa-utah-data-facility.

Castells, Manuel. 1996. The Rise of the Network Society, The Information Age: Economy, Society and Culture Vol. I. Malden, MA; Oxford, UK: Blackwell.

Castells, Manuel. 2000. *The Rise of the Network Society*. 2nd ed. Malden, MA: Blackwell.

Deleuze, Gilles. 1992. Postscript on the Societies of Control. *October* 59 (1): 3–7.

———. 1995. *Negotiations, 1972–1990*. Translated by M. Joughin. New York: Columbia University Press.

Foucault, Michel. 1977. *Discipline and Punish: The Birth of the Prison*. Vintage Books.

———. 2008. *The Birth of Biopolitics: Lectures at the College De France 1978–1979*. Edited by Michel Senellart and translated by Graham Burchell. London: Palgrave Macmillan.

Hardt, Michael. 1998. The Global Society of Control. *Discourse* 20 (3): 139–152.

Ihde, Don. 2002. *Bodies in Technology*. Minneapolis and London: University of Minnesota Press.

Lyon, D. 2001. *Surveillance Society: Monitoring Everyday Life*. Buckingham: Open University Press.

McLuhan, Marshall. 1962. *The Gutenberg Galaxy*. Toronto: University of Toronto Press.

Rheingold, Howard. 1993. *The Virtual Community: Homesteading on the Electronic Frontier*. Reading, MA: Addison-Wesley Pub. Co.

Schumpeter. 2015. Digital Taylorism. *The Economist*, September 12. https://www.economist.com/business/2015/09/10/digitaltaylorism.

Stiegler, Bernard. 2014. *Symbolic Misery—Vol. 1: The Hyperindustrial Age*. London: Polity.

Stone, Jon. 2015. Theresa May's Proposed Spying Law is 'Worse than Scary' United Nations Says. *The Independent*, November 11. http://www.independent.co.uk/news/uk/politics/theresa-mays-proposed-spying-law-is-worse-than-scary-united-nations-says-a6729741.html.

4

The Quandaries of Machinic Subjectivity in Félix Guattari's *Chaosmosis*

Gary Genosko

At the outset of "On the Production of Subjectivity," the opening essay of *Chaosmosis*, Guattari (1995, p. 1) draws on his psychotherapeutic practice and political and cultural commitments in tackling the ticklish matter at hand. His brief Bakhtinian characterization of subjectivity as "plural and polyphonic" that has "no dominant or determinant instance" is framed by a tripartite problem set: (i) current social and political events (Tiananmen Square massacre; pulling back of the Iron Curtain; the first Gulf War); (ii) massive machinic productivity (emphasizing concrete machines such as personal computers, portable music, games, and telephones; magnetic media—home video); and (iii) prominence of ethological and ecological perspectives (arising from chemical, nuclear, and oil disasters in Bhopal, Chernobyl, and Alaska; the emergence of animal rights; sinking of the Rainbow Warrior; signing of the Basel Convention against toxic dumping; and the Rio Convention on biodiversity climate

G. Genosko (✉)
Communication and Digital Media Studies, Faculty of Social Science and Humanities, Oshawa, ON, Canada
e-mail: gary.genosko@ontariotechu.ca

S. S. Das, A. R. Pratihar (eds.), *Technology, Urban Space and the Networked Community*, https://doi.org/10.1007/978-3-030-88809-1_4

change). These three criteria are not fleshed out in any detail, but it is evident that the leading event, the student-led pro-democracy protests in Tiananmen square in June 1989, marked a turning point for Guattari as far as events of collective subjectivation with a global impact were concerned; echoes of which are certainly felt in fall of the Berlin Wall. These global events have complex trajectories that combine progressive and retrogressive desires under the lure of "the Yankee way of subjectivation" (Guattari 1995, p. 3), which represents a shocking wake-up call for those movements modelling change on its values and versions of freedom imposed by mass media and military might.

Capitalist subjectivation was already a major preoccupation for Guattari during the 1980s as it mutated in two forms, on the one hand, towards a more integrated and systematic oppression, combined with immaterial production and semioticization of power. Subjectivity is the foreground against the background of new forms of capitalist valourization, yet this is too simplistic, and must accommodate Guattari's sense that not only are traditional academic disciplines ill-equipped to deal with these new situations, but the "contemporary subjective cocktail" (1995, p. 4) that impresses itself upon him most forcefully pulls simultaneously in two directions, one engaging attachments to archaic religious and cultural practices and beliefs and another cultivating attachments to high-tech and scientific developments. Here Guattari relies on his own experiences in Brazil and Japan to write this recipe, and explains how what it says about subjectivity must be understood transversally, that is, across its existential-territorial incarnations and constellations of valuative, social and semiotic, incorporeal universes. The first matter at hand in this paper is to attend to the crossing of the psychological and the socio-semiotic with mediatic and technological in subjectivity. The second concern will be that of the critique of capitalist time that underwrites Guattari's re-creation of the terms of contemporary subjectivation. Third will be a reflection on pathic apprehension and the distinction between discursive and non-discursive fields. My overall goal is to provide an explication of "On the Production of Subjectivity" that respects and clarifies the positions taken by Guattari, but also calls into question some of the looser connections and efforts to integrate a wide variety of philosophical and psychoanalytic material into his theorizing.

The Heart of Human Subjectivity

In *Chaosmosis*, Guattari (1995, p. 4) situates information and communication technologies (ICTs) at the "heart of human subjectivity." He previously presented this relationship in *Schizoanalytic Cartographies* such that subjectivity enters the machine. Occasionally, he puts it in reverse: "we have to ask ourselves if this 'entry' of subjectivity into 'the machine' (like one used to say 'entering into religion') really is all that new" (Guattari 2013a, p. 2). This approach offers another perspective on a post-humanist relationality among actants by invoking the entrant to religious life, perhaps suggesting despite itself a humanist residue within a post-humanist framework of immanence. Traces of this approach are effectively cleared away in *Chaosmosis*. But the machinic elements involved in these mutually imbricated relationships are quickly reinscribed semiotically in a series of heterogeneous components that are at once: attached to linguistic codes operating in social domains and to mass mediatic significations; and stretched beyond meaning and language in a-signifying machinic signaletics. These latter have not been addressed by structuralism, and Guattari names both Julia Kristeva and Jacques Derrida as post-structuralist contributors to the "relative autonomy of this sort of component" (1995, p. 4) without providing any specific examples (i.e., *différance*, *chora*). The mutual emplacement of ICTs and subjectivity, then, serves Guattari's purpose of criticizing linguistic reductionism, in the first instance, and the marginalization of other sign machines, those involved in the search for non-meaning, in the second instance, which he wants to expand towards a much broader conception of semiosis; one that assists him in revalourizing signals across organic examples like vast fungal formations all the way into the techno-faces of analogue and digital technologies in one of his favourite examples, magnetic stripes on plastic cards, and triggering operations in automated processes (i.e., debit card purchases). These kinds of examples serve to shift enunciation away from human subjectivity and, as he put it, to "refuse to accord to human subjectivity an exceptional existential status" (Guattari 2011, p. 159). To be sure, ICTs are not simply concrete machines; after all, a concrete machine gives only the appearance of being a "clearly defined object," an arbitrary "cut-out" (Guattari 2013a, pp. 94–95). So, while I will integrate a few examples of concrete machines like videotape cameras and cassette decks into my account, it would be reductive and dismissive of the concept to isolate concrete examples from the signaletic flows and evolutionary phyla in which they are embedded, as well as the abstract machines that catalyse assemblages.

Guattari's opening move in *Chaosmosis* seems fraught with difficulties. Certainly, by displacing the capitalized *Subject* with the small case subjectivity, he is not radically separating himself from the philosophical tradition, as he claims to retain processual as opposed to essential features from phenomenology, for example, including intentionality, a less than obvious move (1995, p. 22). "The founding instance of intentionality," as he puts it, contrasts with approaches that make the subject the essence of individuation. This move should be puzzling; for the interiority required for a phenomenological investigation entails a re-centred intending *human* ego-subject functioning within interrelated, inter-subjective unities engaged in meaning formation, which seems to fly in the face of a subjectless, emergent subjectivity that has been the hallmark of his theory's critical reception and post-phenomenological setting. Paul Bains (2002, p. 105) underlines the "radical distinction from phenomenological accounts."

Guattari's immediate shift away from the ego pole onto the question of non-individuated or impersonal enunciations, but not to the object pole, either, instead proceeding by the middle ("taking the relation between subject and object by the middle"), as he suggests, perhaps settles some opening essay jitters by outflanking the Husserlian noesis-noema poles, and the "inertial" shut-in subject, and so on. At the same time he has advocated "abandoning transhistorical essences or phenomenological analyses oriented simply around molar ensembles" (Guattari 2011, p. 159). The question that he asks is how intentionality supports an outflanking of the molar ensemble of subject-object and how this allows for subjectivity to find itself.

If ICTs get at the heart of machinic subjectivity, or vice versa, it further becomes hard to see how, if the analysis of the semiotic and economic composition of semiocapitalism and its tributaries of precariousness and debt are correct, this could offer an exit from capital's levelling of value and crushing of polyvocity by "the subjectivity of generalized equivalence" (2011, p. 159), since it seems to lead to further refined versions of the very problem that Guattari wants to challenge, meet, and surmount. How does a "machinic subjectivity of a new type" in the age of the microprocessor keep ahead of evolution along the phylum of digital machines dominated by big tech firms? Guattari's effort to move forward efforts to grasp the processual potentialities of ICTs is tempered by his mixed observations: that "computational … revolutions so far only result in a reinforcement of previous systems of alienation" (2013a, p. 4), and computers and AI assist in mutant creationist thinking grounded in a-signifying semiotics within a post-humanist orientation that favours non-human machinic proto-enunciation.

One of the goals of this chapter is to set out some initial quandaries of the shift to subjectivity as a process and a passage, and explain how Guattari proposes to find his way by proposing an aesthetic paradigm. Indeed, to speak of quandaries may seem unduly melancholy. Although Guattari does not discuss specific artists or movements at length in *Chaosmosis*, he does broadly name identifiable areas such as performance art and underground art. He then makes strong claims about these, such as "it is in underground art that we find some of the most important cells of resistance against the steamroller of capitalistic subjectivity—the subjectivity of one-dimensionality, generalized equivalence, segregation, and deafness to true alterity" (1995, p. 91). When he does mention in passing an artist, namely, Chilean surrealist and abstract expressionist Roberto Matta, the results are somewhat ambiguous. Guattari's attunement to the heterogenetic transformations (either singularizing or homogenizing) of new technologies was in many ways exceptionally attentive. Take, for instance, his reference to computer-aided engineering and design (originally generic, "l'assistance par ordinateur," but specifically in English "computer-aided design"). The assumption about CAD (computer-aided design) is not precisely matched by the presence of CAO (conception assistée par ordinateur), but the assumption is quite reasonable (Guattari 1995, p. 5; Guattari 1992, p. 16). The CAD revolution in 3D software that emerged through the 1980s and heavily influenced industrial design is not discussed by Guattari. Instead, the availability of this software "leads to the production of images opening on to unprecedented plastic

Universes—I am thinking, for example, of Matta's work with the graphic palette" (1995, p. 5). Despite his longstanding interest in architectural forms and diagrams, Matta was not an adopter of CAD; granted, however, that the three-dimensionality of the objects in the pictorial spaces of some of his works suggests the digital objectality of virtual spaces (Guattari 1987). Guattari's point is surely that creative designs utilizing CAD can achieve something on the order of Matta's extraordinary universes, many of which pre-date this technology, yet seem as if they were created with it, or something like it.

Machinic Hypertext

When *Chaosmosis* was originally published in 1992, the world wide web was still quite new, sketched out by Tim Berners-Lee (1989) across the years 1989–1990 in the European Organization for Nuclear Research. Berners-Lee's critique of arborescent hierarchies in search protocols and assignation of a rhizomatic role to hypertext remain consistent with Guattari's thinking. It was Berners-Lee who contributed key elements to the formation of the web, notably his browser that utilized Hypertext Transfer Protocol and the scheme of the Uniform Resource Locator through which a client requests a resource from a remote host. To be sure, precursors like Ted Nelson had developed the idea for hypertext writing decades earlier as a non-sequential ordering system for written fragments, situating the problem in terms of personal project organization, and Nelson himself attributes the first notion of hypertext to the post-war years, and considers Berners-Lee's solution to be a "brilliant simplification" of what he had in mind 25 years earlier that suddenly caught on (Whitehead 1996). Guattari's user perspective dovetails with Berners-Lee's non-proprietary development of fundamental languages that have come to define the web, despite the many corporate retrenchments along the way since then. Guattari's scattered comments suggest an affinity with Nelson's personal retrieval paradigm of "associated trails," converted into group-subject assemblages, raised to the level of Berners-Lee's more embedded institutional perspective.

Imagine moving hypertextually around Guattari's conception of the grid diagram of work rotations at La Borde: scrolling down the vertical list of names on the left; clicking right to find out which activities each would take on, which would take you to the hourly schedule for that week. Imagine a file that counted the frequency of a person's performance of a task, and how many points they accumulated in so doing (at least for the tasks that generated points as not all did so). One could then hotlink to a file that contained a dictionary of local vernacular terms and jargon (*grilleurs*, *pensionnaires*); and request an updated list of the members of the current management teams of monitors, even coefficients of boredom tied to certain activities, and specific spaces, that would later be highly valourized, could be consulted (Cálo 2016).

My point is to offer a modest imaginary of hypertext without much web surfing. This contrasts with the elaboration offered by Brian Massumi (2002, pp. 138–143) where the critical acknowledgement of the limitations of pre-formed hyperlinks, as well as the possibilist trap of the digital itself, including the apprehension of its affectual contours and analogue relays, if followed closely, retroactively imposes both debates and advanced milieux (closed and open architectures) that did not yet have a strong purchase on the browsing localism of early hypertext and its institutional embeddedness. The early challenge is how to hyperlink to your own resources, locally, without web surfing yet entering into it. When you access your own files you are both client and host, your movement lacks qualitative vagueness because change is slow, relays lack Internet scale magnitudes, accumulation is less impinging, and the stakes are more focused and less open. Hypertext links of the sort I'm describing didn't request remote resources, as the links were for documents in the same organization, often on the same server, or in the same file. Early hypermedia browsers circa 1992–1993, with pre-graphical interfaces, didn't really link to much of anything, and thus the earliest manifestations and institutional contextualizations of these technologies are closer to telephone books.

The machinic character of writing itself, exposed in both the paper technologies of filing and the devices of reproduction from typewriter, to microfiche, through videotext systems where content could be dialled-up and displayed locally, leads Nelson to define hypertext in this way: "a body of written or pictorial material interconnected in such a complex way that it could not conveniently be presented or represented on paper. … Such a system could grow indefinitely, gradually including more and more of the world's written knowledge" (Nelson 1965, p. 96). Guattari shows no obvious direct evidence of familiarity with either Nelson or Berners-Lee, or the long-post-war march towards hypertext as he likely knew it. He thought of linkages between databases (textual and audio-visual) in terms of how directly transversal relations could overcome linearity and specialist isolation, encourage transdisciplinary exploration and the blossoming of new collective assemblages of enunciation. Hyperlinks advance transversality but in a dated telematic framework (engaging connectivity among databanks, audio-visual resources like interactive [two-way] videotex, and telecommunications services, either phone or cable television) in a pre-mass state. In a sense, he doesn't require any familiarity for his deployment of the term, similar to that of his multiple uses of "interface," especially later in *Chaosmosis* where the concept must bear considerable weight, since both will result in an augmentation of transferential relations within his ontological categories. But anyone who has studied the grid (*la grille*) at La Borde, its development and phases, elementary components, visualization, how feedback worked, can appreciate just how a local hypertext capacity would have helped everyone move around with the diagram, especially as the number of participants grew. Interfaces, qualified as both machinic and an "infinite play of" them (Guattari 1995, p. 30), serve as important bridging devices between attributes of pathic logic and logic of delimited sets, especially between existential synchronies like complex refrains, and standard time (Coordinated Universal Time or GMT—Greenwich Mean Time). The pathic subjectivity's adjacence is the power to be affected: to be subjected to life. In intermingling with machines, subjectivation transfers and shares its adjacency so that global extrinsic points of reference like space, time, energy, and profit may be in turn displaced for the sake of intrinsic, intensive, and non-linear coordinates of pathic experience: a *non-discursive transitivism* that is in a fraught relation of reciprocal implication with discursivity, for discursivity rests on the very non-discursivity it attempts to marginalize, and non-discursivity requires discursivity in order to make itself discernible. This will be directly sourced from phenomenological psychiatry, and will be taken up later in this paper.

Inspired by Pierre Lévy, Guattari (2013b, p. 113) came to consider machines along the lines of interfaces. He outlined a conception of proto-subjective "animistic" machines open to assemblages with other components of a machinic environment; this ontological pluralism of machines was not satisfied with allopoetic assemblages of non-living components, however, but regained autopoietic features (as in the relations between biological beings but extended to non-living becomings in relation to its own unfolding and with regard to alterities). In so doing Guattari highlighted the role of proto-subjectivity and proto-animate components that can attach themselves to one another and help build hybrid subjectivities.

Thus, both machinic hypertext and machinic interfaces serve similar purposes: on the one hand, to contribute to an expanded field of enunciation in an auto-affirmative way through the generation of connective tissues; and to further transversalize the relations among such entities and their worlds. One could say that interfaces in Guattari's writings are entities he invents to address ontological problems; so, in *The Interface Effect*, Alex Galloway (2012, p. 30) is quite right: "The interface asks a question and, in so doing, suggests an answer." Simon O'Sullivan (2012, pp. 96–97) captures this well in his *On the Production of Subjectivity*, in explaining how Guattari invents entities to interface categories (the question is: how do they relate?); these transversal entities explain just how they relate, hence they answer the question. Whether these are rather *intrafaces* among entitarian relations is probably a matter of scale, within a quadripartite model of the Unconscious at play. Guattari (1995, pp. 29, 31, 111) thought machinic subjectivation could creatively auto-engender itself by means of dynamic links across many divides, which supports regular denials of his own Manicheanism.

Guattari doesn't reflect on the fact that hypertexts or, more broadly, hyperlinks, are coded in advance in a preferred manner around keywords and anchors, because for him these bridge registers, logics, and domains whose relations are often subject to explicit constraints, anyways (2013a, p. 58). Whether it is between subject-object, mind-matter, finite-infinite speeds, or discursive-non-discursive-domains, these are traversed by machinic entities that immanently inhabit and negotiate them towards the achievement of some consistency.

Machinic Subjectivation

First, the machinic definition of subjectivity diversifies the components involved and avoids reinscribing molar alliances; and second, it describes the attributes of the transferential powers of the relations among such components through a resistance to division, an allergy to developmental schemas, and a reluctance to posit purely pre-formed distillations. Guattari (1995, p. 108) elects "trans-subjective" passages of affects, fantasms, ambient distributions, and many other fluid exchanges. Machinism is creative, ethical, and aesthetic.

The placement of ICTs at the heart of subjectivity assists in dislodging the status that individuation typically enjoys in definitions of subjects-as-persons. Hence, his claim that "subjectivity is collective," with the proviso that it is simply irreducible to social determinants, and doubly displaced from persons onto the socius before the person, that is, to pre-verbal, affective exchanges. This makes the production of machinic subjectivation both "non-human and pre-personal" (1995, p. 9). Collective, as Guattari specifies, means multiple: beyond and before the person. And it engages non-human machines at all scales.

Subjectivity produces itself through both human and other-than-human processes, and with regard to diverse references (some cognitive, others mythic, ritualistic, etc.). These references interact like competing maps. The question that interests Guattari is whether existing maps require revision or replacement by new maps. All such maps need to be evaluated and how they have been used is critically reviewed; can some of these be turned towards the future (diagrams) instead of the past (tracings)? Guattari foregrounds the role of "co-management" as opposed to "authority" (referring to the analytic situation) in subjectivation. A so-called dirty word is even proffered—"a dialectical relation" (1995, p. 12) between cartographic impulse and psychological modelization *and* individuals and groups is entertained. This mediation does not appear to move towards sublation. Guattari's writing here displays a countercultural propensity to distrust authority, directed at psychoanalytic schools, a field of conflict in which he lived. And from within it one can say that Guattari seeks some recognition for his own inventions, and he catalogues his renouncements (of the conscious—unconscious dualism that marked his early work on transversality, for instance) and offers his alternative schizoanalytic cartographies not as scientific theories but as artistic practices for the composition of subjectivities.

The coherence of this view is not appreciated until one grasps that for Guattari the *transferences of subjectivation* operate across categories such as leader-follower, author-reader, and producer-consumer, in a robust co-creationism that is not driven by the logical necessity of a Hegelian dialectic; while there can be a shallow, congratulatory side to this engagement in figures of the *prosumer* and its direct connections with the capitalist economy, undoubtedly in the throes of dot-com era futurism, the main interest here is Guattari's two-track explication of subjectivity through psychoanalytic and aesthetic problematics, back and forth from which he moves throughout *Chaosmosis*. Rather than determinations passing into their opposites, to be preserved in a progressively unfolding higher unity, transferences occur. Transference is, Freud insisted, a repetition of a past experience aroused by the treatment into which the present analyst is inserted in place of another person; subsequent modifications as projections onto the analyst, and the process through which these are re-introjected, often give rise to counter-transferences on the part of the analyst who is interpretively hindered by them, until they are resolved. The movement in psychoanalytic interpretation from a negative to a positive understanding of transference and counter-transferential dynamics (of various colorations) is a phenomenon of the divergent emphases by different schools, not one driven by necessity. Guattari (1995, p. 4) also looks askance at such developments when they veer towards scientism.

Guattari (1995, p. 9) attempts a provisional, general definition of subjectivity: "The ensemble of conditions which render possible the emergence of individual and/or collective instances as self-referential existential Territories, adjacent, or in a delimiting relation, to an alterity that is itself subjective." Territorial instantiation has a double reference: to itself and to an alterity, on a stage set by an ensemble of conditions. The relationship between an individual or collective subject and an alterity (taken in a larger sense of situations, families, communities, laws, objects, and non-human agents) is either one of adjacency or based on fixed boundaries (a "relation of delimitation"). For Peter Pál Palbert (2015, p. 186) "the reiterated reclaiming of the opening to alterity does not make reference to an alterity of another subject (such as in the phenomenological conceptions on intersubjectivity), but rather to a more complete alterity, that is the situation itself." Guattari (1995, p. 9) enumerates a number of conditions for subjectivity's production: human linguistic communication; non-human identificatory learning; institutional interactions such as degrees of transversality among different classifications of workers; interfaces in computing; incorporeal value Universes created by artworks. Guattari is going to put the emphasis equally on adjacency relations and delimited sets. The latter, boundary marking, may be accomplished by refrains, as among birds, or by symbolical circumscriptions in ritual communications within human groups (1995, p. 15).

Adjacency recalls the language of Deleuze and Guattari's *Anti-Oedipus*, in which subjectivity is characterized as a "residuum," "appendix," or "spare part." The machine displaces subjectivity from the centre to the periphery, where it has "no fixed identity … defined by the states through which it passes" (1983, p. 20). Subjectivity results from a certain kind of break; it is a residuum. Subjectivity is a machinic process of cuts and connections, neither a point, interior, mind, nor person. It is exterior, flung centrifugally, outside the hallowed centres of sovereign individuation and ego. It exists beyond the acts of a transcendental ego, a constituting subjectivity with the authority to conjure truth, beyond the reach of the thinking/being condition. Instead, sensations exist where thought was once hoped to purely and surely dwell, and with them uncertainty, revisability, and active and passive extensions.

Guattari's vision of co-management of the production of subjectivity is embedded in the analytic encounter in a collective setting, with the provisos noted earlier. Making this possible was for him a worthy yet difficult task of standing simultaneously in analytic and aesthetic domains. To this end he erects a transversal bridge between the analytical part object (*objet a*—little other) and an aesthetic equivalent detachable from a work's significance; these two processes of "autonomisation" allow adjacent parts or fragments to work relatively independently, yet to apply both to creation and consumption, and to pass between them in acts of co-creation. To be autonomous in the language of machinism means to be free from structural determinants, to be autopoietic, that is, to extend to concrete machines in the sense that they "maintain diverse types of relations of alterity" (Guattari 1995, p. 40) and given that alterity is not itself centred on another subject.

Guattari recasts the little other as a machine that doesn't fall or drop out of the symbolic order, but is extracted or detached by an artist or spectator, furthering subjective autonomisation and transferences of subjectivations. Guattari does not explicitly retain the presupposition of the experience of castration (i.e., as in separation, weaning from the breast), but since he turns immediately to the performance of poetry, the voice is perhaps the best example, and a classic one for Lacananians (voice as *objet a*). In Guattari's hands the little object still gets desire going as it becomes a detachable fragment that enables autonomisation through partial enunciation. Guattari's interest is on the side of expression. In moving away from the emptiness of the leftover utterances to semiotically formed phonic substance, not the evacuated sound-image (Saussurean signifier), but verbal activity, its sonority, nuances of articulation, and the embodied *strivings* of verbalization, in other words, on the action of performance, Guattari abandons Lacan for Bakhtin's poetics, without exaggerating his role as there are many possible sources at play. This is illustrative of the emergence of existential impactful aesthetic experience that supports transferences of subjectivation. Maurizio Lazzarato in *Signs and Machines: Capitalism and the Production of Subjectivity* (2014, p. 189) captures Guattari's use of Bakhtin's non-discursive principle of "the feeling of the creative activity of speech" in his emphasis on the speech act over the signifier.

Aesthetic perception in general becomes for Guattari the moving force for the detachment of fragments from the social field (broader than a work or text) that constitute for creators and consumers alike opportunities for partial enunciations that can support subjective autonomisations. This is the privilege Guattari (1995, p. 131) affords to aesthetic perception as it reshapes subjectivation by rupturing ordinary life and disrupting existing categories of capture: "the focus of artistic activity always remains a surplus-value of subjectivity." The invocation of "'invention' itself" (Zepke 2011, p. 213) is irreducible to a specific movement model and wide enough to situate aesthetic experience's role in Guattari's thought: to counter, for instance, semiocapital's attempt to gain a hold over models and maps of subjectivation. It is his antidote to "a social field devastated by capitalist subjectivity" (1995, p. 15).

In order to investigate art and subjectivation, Guattari discusses a specific non-structuralist variant of psychotherapy; one, to be precise, that was developed by his colleague Mony Elkaïm in his family practice. The production of subjectivity within family therapy is machine-assisted within a largely improvised, highly "ludic," psychodramatic role playing, which is videotaped and played back for comments. A therapist can create new modalities of subjectivity by using concrete machines like videotape, even when it is switched off or a palette of colours is restricted, by subtraction. For Guattari (1995, p. 8), "it should be emphasized that the video is always within sight of the therapists. Even when the camera is switched off, they develop the habit of observing certain semiotic manifestations which would escape normal observation."

Why is this example important? Guattari's "first" analytic work with a schizophrenic patient was with a tape recorder. The "Monograph on R.A." in *Psychoanalysis and Transversality* explains that Guattari turned on the tape recorder only at moments when the dialogue stalled or something vexed him. Switching on the device either deviated the dialogue or ended it. Guattari's method regarding this machine consisted in playing back the tapes of the session to his patient with the goal of bringing him to recognize himself in a kind of deterritorialized aural mirror stage, and shift the recording function onto R.A. himself but this time with a notebook: "After a certain time, the tape recorder had conditioned the situation of our dialogue to the point that I almost did not need to turn it on. I abandoned it and in its place, I wrote down the things he said" (Guattari 2015, p. 40). Two different concrete machines both turned off. Two analogue magnetic media with on (record, playback) or off (stop) modes, not exactly digital (whether one can turn off a digital device like a smart phone remains an open question). Guattari relates that in both clinical examples the off position is open to change, the analyst's perception of semiotic sensitivities becomes heightened, the analogue flows into another analogue medium, handwritten notes on paper. Counterintuitively, momentum belongs to the off position, as much or more than to the on position. Depressing the stop or power button with a finger is a momentary analogue point of contact with the extensions of the virtual, as what follows the button click twists into self-separating varieties of transformation away from the concreteness of the machine, escaping from these into an abstract machinic dimension. What occurs here is an inversion, in a sense, in which the abstract drives the concrete, and the virtual delivers potentialities, in which the off mode has a greater openness and activeness than the on mode. Neither mode is trapped in its actuality, but the emphasis here is on the inactual rather than a resting state and how analysts try to channel such potential by entering into its flows of intensity as directly as the situation permits.

In *The Anti-Oedipus Papers*, Guattari (2006, p. 146) refers to the "magico-machinic technique of the tape recorder (less common back then)." Returning to the inversion with which we began, as it is stated, this time, in *The Machinic Unconscious*—"the subject and the machine are inseparable from one another. A degree of subjectivity enters into every material assemblage. And reciprocally, a degree of machinic enslavement enters into every subjective assemblage" (2011, p. 159); the tape recorder here mediating in the assisted auto-assembly of subjectivity and its encounters with alterities, using "assisted" in the strong sense of the to-and-fro of clinical and aesthetic practices that marks the movement of Guattari's thought in *Chaosmosis*. But also relational as Guattari's definition indicated. While Simon O'Sullivan (2012, p. 263) states, "in the aesthetic paradigm we become the authors of our own subjectivities," in the analytic-aesthetic relation, we receive assistance and mediation, and are relayed to other subjectivating processes. If we were phenomenologists, we would say inter-subjective: Edmund Husserl mystically explained, in his *The Crisis of European Sciences*, how "in an amazing fashion [one's own] intentionality reaches into that of the other and vice versa" in an "intentional interpenetration" (Husserl 1970, p. 254) resulting in an internal unity, "an intentional mutual internality." Guattari's retention of intentionality, however, does *not* foreground interhuman constitution by ego-subjects of an interior mutuality among them. His decentring of the human and closing of interiority neither completely foreclose subjective individuation, nor inter- accumulated in machinic assemblages, his machines of subjectivation don't work with Husserlian unified "souls." This repositions intentionality as non-human and machinic; it is the animistic partial-subjectivity of machines that doesn't intend, but partially enunciate. Partial enunciations are accumulated in machinic assemblages of subjectivation and positioned around (before and outside) of subject-object relationships, Guattari (1995, pp. 24–25) explains, adding that his approach to taking intentionality from the middle is to separate it from both its aboutness and its presupposed beforeness, that is, from a subject-object relationship. "Subjective transitivism" (Guattari 1995, p. 25), with its Lacanian and Freudian lineages, entails a fusion of subject and object, as a result.

Does Guattari, in slyly invoking infantile transitivism, while suggesting in passing Lacan's early deployments of it, solve the challenges of transversality between planes and across domains and logics? The answer seems affirmative, yet the examples he gives render it highly specialized, belonging to hypnosis, suggestibility, hysteria, and anthropological assessments of pre-logical thought. Is this a *dialectical* resolution of the permanent challenge of separation by means of a fusion as a selected higher-level resolution of the subject-object opposition? The psychoanalytic-dialectical origins of the term generate a quandary: is the result of such fusion unity or discord? Jacques Lacan (2006, p. 99) credited child psychologist Charlotte Bühler with the conception of transitivism in children, a phenomenon in which one child may strike another child and claim "he/she hit me" or one child may injure herself and another child might begin to cry in response. But, to answer the question, it is neither, for sublation is not the end game, as Guattari quickly confirms that while transitivism can be regained from the psychoanalytic narrative that vectorized subjectivity's non-discursive, fusional attributes, subjectivity "continues to self-actualize through energetico-spatio-temporal coordinates, in the world of language and through multiple mediations" (1995, p. 25). So, nothing is really cancelled and surmounted since discourse (EST coordinates) continues to play a role, despite throwing up all sorts of problems in the form of binary thought, empty referents, linearity, structural coordinates, standard time scales, and domineering molar concepts (Being, Capital, Signifier, Subject).

Our problem here is not mirrors, but tape recorders and video cameras. The resolution of the discursivity of the intentional relation of subject and object in pulling out the middle is Guattari's strained resolution of at least one of his phenomenological attachments.

Guattari admired New York East Village artist David Wojnarowicz's machinic sensibilities. The artist auto-assembled his subjectivity by means of portable tape recorders throughout his life, beginning by using them to collect stories from his Beatnik friend Hubert Huncke, later using tape as a diary to record his dreams. Wojnarowicz even played the tape recorder in a post-punk band Three Teens Kill Four. He slept with his tape recorder close by and used it upon waking; he would nod off as he spoke into the microphone after heavy drug use. His attachment to the machine was intense, yet listening to his own recorded voice caused him to experience discomfort, especially awkward was how he spoke about his mentor and lover, photographer Peter Hujar. Sometimes, the tape would run out. Eventually, long after his death, Wojnarowicz's tapes were transcribed, printed, and bound as his tape journals, bringing to semiological consistency an otherwise looser, multiple-semiotic consistency whose finer features of deterritorialization (effects of drugs, illness, sleep deprivation, changes in voice while drinking, smoking, bad recordings, squeaky spools, tape hiss, unvoiced sections, background sounds) and experimental treatments (ambient recordings made in moving vehicles or while walking) are almost completely lost in the process of privileging well-formedness (substance) for publication (Wojnarowicz 2018). I am neither lamenting the loss of intimacy nor the realization of a possibility, confusing the virtual with this latter; for Guattari, the "lure of intimacy" must be dissolved (2011, p. 70). Turn it off: once in the stop position, the artist's voice is cut off, interrupting the flow of sound waves and their electrical conversion, but without blocking the eventual material flows leading to the inscription apparatus to which it is connected, as well as to later copying exercises, used by editors and publishers, and clinicians like Guattari (who had R.A. copy Kafka's *The Castle* by hand). Further encounters between and across alterities may be mapped within a phylogenetic evolutionary history of magnetic media that saw digitality muscle aside humble analogue devices, as well as ontogenetic becomings like released semiofluxes, distributions of affects, and its ongoing constraints (a need for batteries and a power cord). Being turned off does not arrest becoming: not to end up as e-junk, but to enter an archive, then a gallery, and the new alterities that continue to occur and accrue—and the reterritorializations of the semiological aggressivity of print publishers, of those who would convert tape to digital files, not to mention those critics who would factor themselves into the concrete machine's expanding abstract diagram. The diversity of types of alterity in which any concrete machine is engaged is for Guattari (1995, p. 42) what makes it autopoietic "ipso facto" by that fact itself. Diagramming virtual extensions entails grasping more than individuation: "a more collective machinism without delimited unity, whose autonomy accommodates diverse mediums of alterity" (1995, p. 42).

Semioflux and Capitalist Subjectivity

For Guattari (1995, p. 29), "capital smashes all other modes of valorization." By smashed, read flattened, in a more extreme sense than Marcusean one-dimensionality, even though his sense was polyvalent and applied to distinctions such as high and low, ideal and material, refusal and accommodation (high and mass culture). How does this work? Earlier, however, in *The Three Ecologies*, Guattari valued the pursuit of the idea that value could be conceived of anew outside of general equivalence; "the information and telematics revolutions are supporting new 'stock exchanges' of value" (Guattari 2000, p. 65). ICTs could, he believed, give rise to such new exchanges and he didn't mean the NASDAQ. It is not clear that he would have considered virtual currencies to be much better. The quandaries that we have considered above turn us back to the more explicitly political question of Guattari's position on capitalism and subjectivity.

By the end of the 1980s, a propensity was critically identified and linked to the production of subjectivity within the rise of info-capitalism in the wake of widespread computerization and emergence of network society. Accordingly, subjectivity was produced through knowledge generation and dissemination; by means of the cultural industries; and sensibility and sociability were refined and reinforced through juridical standards and techno-scientific findings, in ways that were co-constitutive with economic mechanisms (Guattari 2000, pp. 47–49). The idea that capitalism manufactured specific kinds of subjectivity over the life cycle was conducted through a semiotic lens. Capitalism had a specific semiotic focus on the conformity of otherwise singular infra-personal attributes to dominant models of social identity in a process of domestication that limited the terms of the relationship. The effort to typologize and historicize subjective types—serial subjects or salaried employees; precarious subjects consisting of unemployed or marginally employed persons; and elite subjects of executives and wealthy families—gave way to a more acute question about the modalities of subjectivation within the information revolution that demanded a highly cognitive form of labour coupled with an imperative for communication and its effects on class composition (2000 pp. 61–63). Namely, the potential for a middle classification of working-class subjectivity as opposed to an earlier serialization of the working class.

Semiocapitalism, a term coined by Franco Berardi, was introduced to describe a form of capitalism that operated semiotically. Berardi generated his theory during the dot-com bubble of the second half of the 1990s, and attempted to identify a non-localizable process of generating surplus value neither with labour nor with workplaces in traditional industrial senses of these terms. He combined the fragmentation of labour dominated by a globalizing machinic phylum with the capacity of the digital network to activate high-tech labour at any time and from any nodal point in the network for the combination and recombination of semiotic fragments, thus framing an info-commodity. Because high-tech labour tended to blur the work/leisure distinction, and acquire a highly personal tone, the cycle of production was not amenable to work-time but to exact moments of demand.

The dot-com era of informatically machined subjectivity focused on cognitive entrepreneurialism as exemplary of an involution that self-exploited one's intellectual capacities in heightening the production of subjectivity and simultaneously one's own exploitation. As Lazzarato (2007) puts it, self-entrepreneurialism made one both master and slave, the subject of enunciation [subject produced by discourse] and the subject of the statement [first person] at the same time and, in this way, fundamentally embattled.

Berardi and Lazzarato's approaches dovetail in the Guattarian hypothesis that "semio-capital is capital-flux that coagulates in semiotic artefacts without materializing itself" (Berardi 2007, p. 76). Here Berardi's descriptions of semiocapitalist production and Lazzarato's immaterial labour hypothesis join together in a vision of a thoroughly machined form of capitalist subjectivity: a human-machine relay exists between the network and the independent contract labourer, a node in an internal process of communication in which infra-personal creative capacities are exploited upon demand. The immaterial labour hypothesis, which spoke both to the labour process and to the product of labour, was a tendency and not an absolute. In Berardi's case, semiocapitalist production closed the gap between execution and innovation and deferred the moment of materialization by prioritizing productive labour's work with simulations that may later become actualized in matter, like video game cartridges. Labour is pulverized into precarity and machinically engages with a semioticized production process that is largely dematerialized (Berardi 2008, p. 20).

What is Guattari's response to the subjectivity of semiocapital? In the era of Amazon Mechanical Turk workers and the atomization of labour time, is labour time still calculable in recognizable temporal units like hours or even minutes? The divergence of value and productive time as a measurable amount in social media suggests that "labour ever more creates value in ways that are poorly related to quanta of time," perhaps not completely, since in some commodities labour time remains congealed as a "minor parameter" (Arvidsson and Colleoni 2012, p. 20).

The varieties of capitalist subjectivity discussed above do not exhaust the processes of subjectivation that result from a push beyond identity, defined and constrained by sociological and philosophical markers. Subjectivity is not condemned to a spiral of capitalist self-elaboration. Guattari insists that "the future of subjectivity is not to live indefinitely under the regime of self-withdrawal, of mass mediatic infantilization, of ignorance of difference and alterity—both on the human and the cosmic register" (1995, p. 133). He advances refrains as ways of "keeping time," and he draws on historical and contemporary examples in so doing, citing detached objects like a symptom, which "through its own repetitiveness functions like an existential refrain" (1995, p. 26). If time under semiocapitalism has become a problem, then it is one that subjectivity must wrestle with reference to new maps and, as Guattari recommends, "catalyzing modules" (1995, p. 16) that evoke new incorporeal Universes of coordinates adequate to highly chaotic existential Territories. Despite capitalistic time's flattening, as Guattari puts it, and it is so pulverized that it damages existing standards of temporal measure—what is called into question is "universal time" like Universal Coordinated Time projected in the manner that engineer Sandford Fleming helped, in the late nineteenth century, to devise as a "cosmic day" projected (a "hypothetical projection," is to use Guattari's terms, 1995, p. 16) like a "dial-plate" over the earth at 15 degree intervals around the globe from the zero meridian—new partial-modules of temporalization may appear, indexed not to external coordinates and boundaries, but to intensive and complex temporalities that are not represented but not detected at all until they are being engendered. Guattari makes much of the "creative moment" as a focal point for the release of temporalities out of which refrains arise and existential rhythms can be laid down and the Universes thus unfurled are newly constellated. These are not externally coordinated Universes with geotags, time stamps, and the latest scheduling software, but "they are given in the creative moment, like a haecceity freed from discursive time—nuclei of eternity lodged between instants" (1995, p. 17). Not yet formed, and freed from exterior coordinates of the EST type, and singular, contingent—an incomparable moment. Such catalysing *temporal* modules have the power to plunge one down and carry one aloft (1995, p. 16). These are Guattarian weapons on the battlefield against capital, captured nicely by O'Sullivan in his effort to explain the importance of a kairotic orientation, which he splits between a regrounding in intention (his use here is distributed among "attention" and "vascillation"), but also, renders diagrammatically as an oblique line; either way, it is in the service of a "generative temporality" (2012, p. 123). O'Sullivan's effort to mutually imbricate Negri's sense of kairos and Guattari's turn to intensive time operates across philosophical systems through similar "registers of subjectivity." This is the moment when Guattari's orientation shifts from ecosophy to pathosophy. I will return to this in the following sections of this chapter.

Pathic Time

Guattari's preoccupation with time is a feature of his turn to pathosophy, in the sense used by Von Weizsäcker (1958, p. 220) to describe the "power of being affected" (Palbert 2015, p. 119); to submit to life by a suffering that is revealed most intensely through periods of crisis. Crisis carries potentialities for subjectivation and new ways of keeping time as objective determinations recede.

Guattari's discussion of how subjectivation keeps time existentially gathers an eclectic set of examples, including the sonority of poetic phrases, bird songs, and ritual performances, and sound expressions of social and governmental divisions (Egyptian and Greek *nomes*). These refrains are turned towards the important principle of crossing subjective thresholds by Guattari (1995, p. 16) in order to focus on their transversal attributes as "intensive nuclei of temporalization."

The example of the complex refrain of watching television follows and it is much more well-developed than those preceding it. The complexity of the refrain owes a good deal to his layering of affect over a viewer who sits at the intersection of componential flows: hypnotic qualities of the screen (perceptual); captivating narrative content (programme); local distractions (a whistling kettle); and engaging daydreams. Guattari puts himself into this example as a porous, polyphonic subjective process, and the question of his unicity is always "relative" to the degree to which the television screen fixes him, and mixes his identity with the talking head on the broadcast, attracted by a detachable *leitmotiv*, in other words, a recurring melody that accompanies the appearance of a character; that detaching is here done by the composer and scenographer who matches sound and vision. The leitmotiv instals itself, Guattari insists, and is the dominant "attractor" in an otherwise chaotic flux of components jockeying for position. This existential refrain is part of a process of subjective deterritorialization. Musical leitmotivs derived from Wagnerian music abound in popular film as well as in television series. These theme songs and sound effects serve as refrains that shift them into the territory of the viewing experience in all of its heterogeneity. In this way they become complex, mutual imbrications. As long as the refrain does not harden, and thus obsessively overcode by reducing heterogeneity, components of subjectivation will continue to pass through the territory, and there is no reason to assume, for Guattari, that subjectivation would be threatened by being pushed into either an implosive posture, or into an explosive, quite delirious, uncontrolled centrifugalization of components.

Pathic, intensive time is a specific feature of schizoanalysis: "time is not something to be endured; it is activated, orientated, the object of qualitative change" (1995, p. 18). Pragmatically, the capacity to "activate" a complex refrain when an opportunity presents itself (a patient makes a passing remark, "I feel like learning word processing") requires that the analyst take it up as a "potential bearer of new constellations of Universes of reference" and the "construction of subjectities" (1995, p. 18). The clinical advice here is to avoid a non-interventionist posture and to take risks, removing the necessity of indexing any refrain to a pre-established hermeneutic. The schizoanalyst does not, in addition, perhaps like Lacan, feel it necessary to suffer the entire analytical hour. While this is undoubtedly the case, it does not respect the fundamental distinction that Guattari has borrowed, and adapted to his own ends, from phenomenological psychiatry between intensive/extensive; subjective/objective. *Immanent lived time*, to borrow Erwin Straus' (2012, pp. 207–208) terms, proceeds towards the future as a feature of personal growth; it is neither measurable nor does it pass in quite the same way as universal time. Guattari is not much interested in grounded subjective temporal experience in biological events. However, he embraces the phenomenological observation that subjective processes engage both kinds of time. In endogenous depression, Straus argues, when there is no longer a future direction—it comes to a standstill—since an "alteration of the potential action and its effects" has occurred, changing the past as well. For Guattari, the clinical situation he invokes in passing is of a patient who has hit the wall, remains stuck, spins in circles. A singularity—"a rupture of sense, a cut, a fragmentation, the detachment of a semiotic content" (1995, p. 18)—that occurs and is exploited analytically reinvigorates the future, it opens it up once more to qualitative change; Guattari also hints that the past may be once again a supportive structure, and not reassert itself in a compulsive framework. Rather, the past may be *renewed-revisited-regained* to gather his terms together; even so, the future may appear threatening, but it is still the future and a propitious turn such as the creative Guattarian cut may dial back the threat, while bearing a new kind of opening: to "originate mutant nuclei of subjectivation." What Guattari does here is rupture the phenomenological perspective on development and growth—time as growth and *becoming* (Von Gebsattel 2012, p. 217) for a personality in formation, more or less interlocked with social and natural cycles. But this is too personological for his schema, thus Guattari turns to Daniel Stern for a more complex conception of emergent subjectivity, one free from psychogenesis, universal marks of development, and the reflux of earlier fixations and, instead, deploys simultaneously existing emergent shifts beginning with the non-conscious self that is transitivist and amodal. In doing so, Guattari inverts the role of rhythm, as it is no longer that which is primordial and disturbed by alterations brought on by psychical disturbances such as melancholy; he turns it into a contributing factor in complexifying subjectivation. This, too, is precarious, he underlines, citing the threat to dissolve existing existential Territories in the move to online relationships (i.e., video conferencing; 1995, p. 19). "Mutant nuclei" play the role of initiating the "overthrow" of redundancies (i.e., subjective types determined by social semiological assemblages, Guattari 2011, p. 209) either by upsetting their existing classification, or by seizing on one redundancy and conferring on it an a-signifying line of force that is characterized as a "mutant rhythmic impetus of a temporalization able to hold together the heterogeneous components of a new existential edifice" (1995, p. 20).

Pathic Knowledge

My earlier remarks on the examples of transitivism (infantile) borrowed from psychoanalysis by Guattari noted only in passing the anthropological referent, Lucien Lévi-Bruhl's *participation mystique*, that furthers the importance of subject-object fusion. For Lévi-Bruhl, so-called primitive mentality was both pre-logical and mystical, the latter parallel to Guattari's real power of the imperceptible in a collective, affective, experience. He writes: "there was what they call 'participation,' a collective subjectivity investing a certain type of object, and putting itself in the position of an existential group nucleus" (1995, p. 25). Mystical participation enables an imaginary transitivism, as Lévi-Bruhl (1985, pp. 76–77) observed: "In the collective representations of primitive mentality, object, beings, phenomena can be, though in a way incomprehensible to us, both themselves and something other than themselves ... they give forth and they receive mystical powers, virtues, qualities, influences, which make themselves felt outside, without ceasing to remain where they are." In the course of regaining this anthropological version of transitivism alongside psychoanalytic examples, Guattari in addition referred to the fundamental role of intentionality in phenomenology:

> It is in this zone of intersection that subject and object fuse and establish their foundations. It concerns a given that phenomenologists have addressed when they demonstrate that intentionality is inseparable from its object and involves a 'before' in the discursive subject-object relation. (1995, p. 25)

As noted earlier, Guattari referred to intentionality as foundational and optimally approached when the subject-object relationship is taken by the middle. Here, his approach is to build up a diverse battery of transitivist examples from across disciplines, including the aboutness of intentionality (directed towards something) and, although his phraseology is a bit unclear here (in English, "before" is marked as a translation of an awkward phrase *d'un en-deçà*), the antecedent in question grants priority not to the ego and object poles but to their inseparability (Husserl 1970, p. 171). The three headings or aspects of intentionality are: "ego-cogitatio-cogitata." It is this wholeness that links the felt participation that unifies all things for Lévi-Bruhl and the general notion of intentionality in phenomenology. These might seem incompatible because of Husserl's disdain for anthropologism as a variation on naturalism, yet under the influence of Lévi-Bruhl Husserl rehabilitated anthropological motifs such as the importance of animism as a naturalistic attitude towards the world not as a dead object (objectivism). Husserl identified European civilization as a spiritual unity that is contextualized as recoverable from the failure of rationalism, and that the relatively meagre insights available about the histories of primitive peoples were actually a benefit to grasping their purposive lives and holistic world-representations. In their study of Husserl's letter to Lévi-Bruhl, Dermot Moran, and Lukas Steinacher point out that "it is clear that Lévy-Bruhl's conception of the primitive mentality had an enormous imaginative influence on Husserl's thinking concerning the development of cultural forms" (2008, p. 18). Guattari does not indicate that he knew of the letter in question. However, consideration of it helps to accommodate the disparate examples he provides, specifically when it comes to the contrast between modes of knowing hostile to rationalism. Ultimately, his undertaking is towards the articulation of the apprehension of the world through pathic subjectivation, which is non-discursive and "before the subject-object relation" (1995, p. 25). Mystical participation, animism, psychosis, infantile transitivism, all assist in Guattari's effort to ensure that pathic experience is irreducible to the coordinates of energy-space-time through which it nonetheless self-actualizes in multiple-semiotic modalities, including language. Guattari (1995, pp. 25–26) continues: "but what allows us to grasp the force involved in the production of subjectivity is the apprehension through it of a pseudo-discursivity, a detournement of discursivity, which installs itself at the foundation of the subject-object relation, in a subjective pseudo-mediation."

Guattari's strategies regarding how he takes the subject-object relation—either in the middle or before it via pathic knowledge or even paradoxically yet foundationally a rerouted discursivity—are in the service of settling what he once called "the ancient Manichean struggle opposing a pure subject and a pure, amorphous matter" (2011, p. 149). Guattari recasts this struggle in terms of a distinction between discursive and non-discursive logics, which is itself a recapitulation of the distinction in Von Weizsäcker (1958, p. 220) between ontic and pathic knowledge. Discursive logic is at worst supportive of capitalist subjectivation, but most of the time it serves rational pursuits and tends to crowd out or bracket it: "science is constructed by bracketing these factors of [pathic] subjectivation" (1995, p. 26). Pathic non-discursivity belongs to intensivity, transitivism, but has no extrinsic reference points, except in the process of existensifying itself in utilizing EST coordinates as noted earlier (discursive links can be utilized in the passage from catastrophe to creation). In other words, pathic process needs some discursive coordination. Guattari considers all subjectivation to be at root pathic, even though this is "overshadowed" and "circumvented" by discursive referents (Capital, Signifier, Being). He can then claim that discursivity rests on non-discursive foundations, given that any subjectivation, even its worst excesses, is at root pathic. Hence, pathosophy becomes his route into chaosmosis. In subjectivation there may be grasped a pseudo-discursivity, a rerouting of discursivity, a "subjective pseudo-mediation" (1995, p. 26). Guattari draws attention to the tendency to impose modes of discursivity and human communication on otherwise proto-subjective phenomena, variously called by him unconscious or partial-subjectivity, proto-subjectivity, machinic unconscious, machinic creationism: "What would a machinic freedom be? Everything here is a question of degree, of imperceptible threshold crossings. Some modes of discursivity, deliberation, and choice exist that do not rest on a signifying discourse between listeners and speakers" (2011, pp. 156). Such freedom means a heterogeneous semiotics, multiple subjective types, undisciplined intensities, an infinite number of components, diversities, and complexities, and thus surprising assemblages.

Another approach to overcoming the sway of binarization may be seen in Guattari's "detour" through the glossematics of Louis Hjelmslev. This "detour" was announced in *The Machinic Unconscious* as a respite from structuralism and semiology, despite the obvious coincidence of the expression and content planes with the signifier and signified of De Saussure's linguistics (2011, p. 41). However, Guattari's approach is to interpret the contraction of the two functives of expression and content, and their relational and reciprocal definition ("mutual solidarity"), in order to delinguistify semiotics. To this end he turns to Hjelmslev's explanation of the relationship between the tripartite division of form-substance-purport (unformed matter) that is distributed across the expression (phonic) and content (conceptual) planes. Exposing the formalism of this Hjelmslevian operation in which scientifically formed matter is semiotically formed, the expression purport appears as featureless sounds and amorphous thoughts. These are formed into expression-substance and content-substance by the projection of form onto purport by pronunciation in a particular language; the expression-form consisting of phonemes and the content-form consisting of selective arbitrary relationships linking sounds to meanings in a given language. Guattari does two things here: he identifies a "common form" across the phonemic and syntagmatic expression plane and the semantic content plane that is an abstract machine engaged in deterritorialization (1995, p. 23). This furthers his strategy of delinguistifying the expression-content relationship and the "integration into enunciative assemblages [of] an indefinite number of substances of expression, such as biological codings or organizational forms belonging to the socius … the question of enunciative substance should also be outside the framework of Hjelmslev's tripartite division" (1995, p. 24). Expression-substance could then take on new and more diverse productions on the basis of technological, aesthetic or non-human materials, and these would be transversally linked and machinically assembled. What he proposes is to displace the form-substance relationship from its centrality and to limit its conceptual priority, and, in this way, decouple the relationship. So, enunciative substances may be composed not from an existing form imposed from the outside or above it on an undivided matter, but rather, by means of both marshalling deterritorializing processes already at work in the matter and available for transformation, at least potentially, for an abstract machine (Berressem 2018, p. 136). When these processes are assembled as substances of machinic subjectivity, this takes place *before* the specification of a subject and object, and are installed in their fusional zone.

Conclusion

The quandaries of machinic subjectivity as an auto-affirming process discussed above show how Guattari sought to reckon with the growing influence over life of ICTs. His efforts to address ontological challenges by extracting key concepts such as hypertext and interface and deploying them as devices assisting in the passages across ontological planes and domains left him on an uneasy ground as he enlisted phenomenological concepts that required heavy emendation. His phenomenological musings were accompanied by a brief dalliance, via the early Lacan, with dialectical thought. I focused my efforts on working through some of the idiosyncratic interpretations he offered, regarding intentionality taken from the middle, the animistic sense of proto-subjective and proto-enunciative non-human components of subjectivity, and by putting a constraining frame around the use of hypertext that was consistent with the early days of the web in the 1990s.

I introduced the importance of certain concrete machines, the video camera that Elkaïm used in his family practice, and the tape recorders that Guattari and Wojnarowicz used in their respective clinical and artistic practices, in order to show how he oscillates between therapy and art, and searched for a way to express machinic autopoiesis—"there is *more* in the machine than simple interactions and systemic retroactions" (2013b, p. 287)—within an unfolding diagram of diverse alterities.

The opening essay of *Chaosmosis* clears the ground for the pluralization of the conception of *autopoiesis* in a full-blown *heteropoiesis* and multiplication of different types of intersecting relations of alterity. Guattari uses the challenges posed by capitalist subjectivity as a constant reminder of what is at stake in finding solutions to its de-complexifying and de-singularizing "mutilation of mentality." Meeting the challenge of the damages wrought by capitalist valourization is the political impetus for his effort to creatively reinvent subjectivity.

Although Guattari does not fully explicate his four-part ontological diagram of Phylum-Flux-Universe-Territory until the third essay in *Chaosmosis*, he develops his earlier explicatory diagrams of this fourfold metamodel from *Schizoanalytic Cartographies*. The key distinction in the first essay is between actualized discursive (including Phylum and Flux) and virtual non-discursive registers (including Universe and Territory). The emphasis he places on the role of interfaces between registers gives an explicit role to complex refrains. These may be "desperately," as Guattari puts it, fulfilled by capitalized Signifier, Capital, or Being within a reactively discursive set of extensional coordinates, yet he emphasizes the incorporeal Universes which may lead to embodiments in existential Territories, not in virtue of coordinates, but through the support of intensive ordinates: "Pathic expression is not placed in a relation of discursive succession in order to situate the object on the basis of a clearly delimited referent" (1995, p. 30). In the pathic journey of experimentation (Palbert 2015, p. 179), discourse is rerouted, stripped of its sense, releasing a-signifying fragments, and not objectively framed and coordinated in space and time.

Guattari's position on phenomenological themes shifts by the end of the first essay of *Chaosmosis*. By identifying the Heideggerian distinction between Being and beings as a "dualist opposition," Guattari (1995, pp. 28–29) distances himself from "Being with a capital B." For Guattari, dualist thinking is a reductive approach he identified with structuralist categories, ancient Manicheanism, and a limitation on multiplicity. Capitalisation is an exclusivist type of valourization Guattari indexed on the actions of flattening, reductive ontologizing, normalizing, and a general emptiness ("empty referent"). Indeed, he states: "Being is like an imprisonment which blinds us to the richness and multivalence of Universes of value which, nevertheless, proliferate under our noses" (1995, p. 29). Guattari positions Universes (not Universals)—virtual, incorporeal, non-discursive, highly deterritorialized, self-referencing—against capitalized and extensional referents. Guattari does not in this gesture expel phenomenology altogether from his theoretical edifice. Rather, pathic apprehension of chaosmotic immanence remains enunciative, and thus to some degree intentional. While Guattari (2013a, p. 257) was not averse to quoting Heidegger constructively, extracting from the Being-being pairing the fragility of existential functions, his effort in the final pages of "On the Production of Subjectivity" is aimed at bringing to a close an analysis of the lack of closedness and pre-formedness of enunciative assemblages, with their "infinite play of interfaces" (1995, p. 30), understood to be plugged into a vast, planetary mechanosphere. In this observation "there is no Being already installed throughout temporality," Guattari reiterates that Being is not the sort of fixed referent available at any point along a temporal line, the understanding of which is the condition for any analysis of beings. This is a double rejection—of ontological difference as just another dualism and of succession (linearity) as a residue of discursivity. Time is not here an extensional coordinate, an empty container, Guattari claims, that can be filled with content, and may be homogenous. Rather, temporalisation is for Guattari a phenomenon of "machinic synchrony": heterogenetic interfaces of a machinic and not only technological character do not require pre-formed components, an object, its representation, and an underlying enunciative subject. The fourfold diagram of metamodelization is, ultimately, introduced in the final lines of this chapter, and it is fourfold to ward against falling back into dualism, by the definition of the fourth field, Universes, as fundamentally an "nth term: it is the opening onto multiplicity" (1995, p. 31). This *three plus n* solution stands in favour of virtuality, not yet formed, in its pure difference.

References

Arvidsson, A., and E. Colleoni. 2012. Value in Informational Capitalism and on the Internet. *The Information Society* 28 (3): 135–150.
Bains, Paul. 2002. Subjectless Subjectivities. In *A Shock to Thought*, ed. Brian Massumi, 101–116. London and New York: Routledge.
Berardi, Franco. 2007. Schizo Economy. *Substance* 36 (1): 75–85.
———. 2008. *Félix Guattari: Thought, Friendship and Visionary Cartography*. Translated by G. Mecchia and C. Stivale. London: Palgrave Macmillan.
Berners-Lee, Tim. 1989. Information Management: A Proposal. *CERN Hypertext Proposal*, TBL900620 (March), 1–21.
Berressem, Hanjo. 2018. 'Degrees of Freedom': Félix Guattari's *Schizoanalytic Cartographies*. In *Schizoanalysis and Ecosophy*, ed. Constantin V. Boundas, 128–148. London: Bloomsbury.
Cálo, Susanna. 2016. *The Grid*. Seminar paper presented at Anthropocene Curriculum & Campus, House of World Cultures (HKW) (Axiomatic Earth, Technosphere Issue). https://www.anthropocene-curriculum.org/contribution/the-grid. Accessed December 2, 2020.
Deleuze, Gilles, and Félix Guattari. 1983. *Anti-Oedipus*. New York: Viking.
Galloway, Alex. 2012. *The Interface Effect*. London: Polity.
von Gebsattel, Viktor. 2012. Compulsive Thought Relating to Time in Melancholia. In *The Maudsley Reader in Phenomenological Psychiatry*, ed. M.R. Broome et al., 214–218. Cambridge: Cambridge University Press.
Guattari, Félix. 1987. L'oestrus: entretien avec Matta. *Chimères* 3: 1–28.
———. 1992. *Chaosmose*. Paris: Galilée.
———. 1995. *Chaosmosis*. Translated by P. Bains and J. Pefanis. Bloomington: Indiana University Press.
———. 2000. *The Three Ecologies*. Translated by I. Pindar and P. Sutton. London: Athlone.

———. 2006. *The Anti-Oedipus Papers*. Translated by K. Gotman. Los Angeles: Semiotext(e).

———. 2011. *The Machinic Unconscious*. Translated by T. Adkins. Los Angeles: Semiotext(e).

———. 2013a. *Schizoanalytic Cartographies*. Translated by A. Goffey. London: Bloomsbury.

———. 2013b. *Qu'est-ce que l'écosophie?* Paris: Lignes/IMEC.

———. 2015. *Psychoanalysis and Transversality*. Translated by Ames Hodges. Los Angeles: Semiotext(e).

Husserl, Edmund. 1970. *The Crisis of European Sciences and Transcendental Phenomenology*. Translated by D. Carr. Northwestern University Press.

Lacan, Jacques. 2006. The Mirror Stage as Formative of the I Function. In *Ecrits: The First Complete Edition in English*, trans. by B. Fink. New York: W.W. Norton, pp. 75–81.

Lazzarato, Maurizio. 2007. Strategies of the Political Entrepreneur. *Substance* 36 (1): 87–97.

———. 2014. *Signs and Machines: Capitalism and the Production of Subjectivity*. Translated by J.D. Jordan. Los Angeles: Semiotext(e).

Lévi-Bruhl, Lucien. 1985. *How Natives Think*. Translated by L.A. Clare. Princeton: Princeton University Press.

Massumi, Brian. 2002. *Parables for the Virtual*. Durham: Duke University Press.

Moran, Dermot, and Lukas Steinacher. 2008. Husserl's Letter to Lévi-Bruhl: Introduction. *The New Yearbook for Phenomenology and Phenomenological Philosophy* VIII: 325–347.

Nelson, Ted. 1965. *A File Structure for the Complex, The Changing and the Indeterminate*. Proceedings of the ACM 20th National Conference. https://csis.pace.edu/~marchese/CS835/Lec3/nelson.pdf. Accessed December 2, 2020.

O'Sullivan, Simon. 2012. *On the Production of Subjectivity*. London: Palgrave Macmillan.

Palbert, Peter Pál. 2015. *Cartography of Exhaustion*. Minneapolis: Univocal.

Straus, Erwin. 2012. The experience of time in endogenous depression and in the psychopathic depressive state. In *The Maudsley Reader in Phenomenological Psychiatry*, ed. M.R. Broome et al., 207–214. Cambridge: Cambridge University Press.

Weizsäcker, Victor von. 1958. *Le Cycle de la Structure*. Translated by Michel Foucault. Bruges: Desclée de Brouwer.

Whitehead, Jim. 1996. Orality and Hypertext: An Interview with Ted Nelson. *The Cyberspace Report*. https://www.ics.uci.edu/~ejw/csr/nelson_pg.html. Accessed December 2, 2020.

Wojnarowicz, David. 2018. *Weight of the Earth: The Tape Journals*. Pasadena: Semiotext(e).

Zepke, Stephen. 2011. From Aesthetic Autonomy to Autonomist Aesthetics: Art and Life in Guattari. In *The Guattari Effect*, ed. E. Alliez and A. Goffey, 205–219. London: Continuum.

5

Fabulation in a Time of Algorithmic Ecology: Making the Future Possible Again

Chantelle Gray and Aragorn Eloff

Introduction: In the Undergrowth of Algorithmic Ecologies

> *The contemporary world—tied up in its ecological, demographic and urban impasses—is incapable of absorbing, in a way that is compatible with the interests of humanity, the extraordinary technico-scientific mutations which shake it. It is locked in a vertiginous race towards ruin or radical renewal. All the bearings—economic, social, political, moral, traditional—break down one after the other. It has become imperative to recast the axes of values, the fundamental finalities of human relations and productive activity.* (Guattari 1995, 91)

> *Considering subjectivity from the point of view of its production does not imply any return to traditional systems of binary determination—material infrastructure/ideological superstructure.* (Guattari 1995, 1)

C. Gray (✉) • A. Eloff
School of Philosophy at North-West University, Potchefstroom, South Africa
e-mail: aragorn@newframe.com

S. S. Das, A. R. Pratihar (eds.), *Technology, Urban Space and the Networked Community*, https://doi.org/10.1007/978-3-030-88809-1_5

In his 2008 article, "The End of Theory: The Data Deluge Makes the Scientific Method Obsolete", *Wired* editor Chris Anderson notoriously argues that the era of the scientific method built around testable hypotheses is fast-becoming obsolete due to the advent of "petabyte-scale" data accumulation and analytics. Observing that tech companies like Google treat this "massive corpus as a laboratory of the human condition" (Anderson 2008), Anderson polemically implores us to forget every theory of human behaviour, from linguistics and sociology to ontology and psychology. "With enough data", he avers, "the numbers speak for themselves" (ibid.), and statistical correlation renders superfluous any theorisation of causality. The Californian Ideology (Barbrook and Cameron 1996)[1] implicit in this overstatement of the benefits of data-driven research has been challenged by several commentators (see, e.g., Pigliucci 2009 and Mazzocchi 2015) and is patently specious. The appeal made by Silicon Valley zealots to 'pay no attention to the man behind the curtain' relies on a mythical, and sometimes even eschatologically feverish, conception of science as a purely objective pursuit, as well as a facile conception of human thought and behaviour—indeed, subjectivity—wherein we can be comprehensively understood via the purely quantitative approaches of Big Data analyses and the statistical aggregation of deep learning systems. As Louise Amoore underscores in *Cloud Ethics*, this ideology confronts our "fallible, intractable, fraught political world with a curious kind of infallibility. In the cloud, the promise is that everything can be rendered tractable, all political difficulty and uncertainty nonetheless actionable" (Amoore 2020, 55). This dream of complete mathematical and technological control over reality—of "a kind of atlas of clouds for the ineffable, a condensed trace of the trajectories of our future lives with one another" (ibid.)—is hardly new. What legal theorist Antoinette Rouvroy refers to as "algorithmic governmentality"[2] (Rouvroy and

[1] The Californian Ideology refers to a 1995 essay by the same name written by Richard Barbrook and Andy Cameron, first published by *Mute Magazine* and reprinted in 1996 by *Science as Culture*. In it, they present a critique of what they term "dotcom neoliberalism", arguing that the concomitant rise of American neoliberalism and the spread of networking technologies in Silicon Valley in the 1990s are strongly correlated with a form of optimistic technological determinism.

[2] Here, Rouvroy is inflecting Foucault's understanding of governmentality towards the digital, where governmentality refers to those practices of sovereign power that influence the behaviour of

Stiegler 2016, 6), by which she means "the increasingly statistical governance of the 'real' ensuing from a convergence of contemporary technological and socio-political evolutions" (Rouvroy 2011, 119), can be traced at least as far back as first-order cybernetics (Wiener 1965, 12) and, before that, to Leibniz's *calculus ratiocinator* (cf. Couturat 1901). What makes the contemporary problem novel, however, is that the latest technologies, for example cloud computing, practically unlimited data storage, high-speed global communications networks and, most importantly, machine learning—and here we mean specifically new forms of connectionist 'artificial intelligence' that, unlike older symbolic AI models, rely on multi-layer artificial neural networks that are assumed to represent the biology of human cognitive structure—have powerfully exacerbated the quest for (and assumptions of) 'Algorithmic Supremacy'. As digital technology theorist Dan McQuillan argues, data-driven modelling via statistical induction is assumed to bear inherent significance, but this approach, which McQuillan describes as a form of "machinic Neoplatonism" (McQuillan 2018), tends to entirely elide the broader subjective and inter-subjective contexts within which analyses and modelling unfold and is thus, via this unwarranted delimitation, able to present itself as operating with a level of mathematical objectivity it simply does not possess.

This creates several insidious problems. For one, as Bernard Steigler observes in *The Age of Disruption*, this deferral to the 'superiority' of algorithmic reason legitimises "the systematic exploitation and physical reticulation of interindividual and transindividual relations" in the service of the data economy (Stiegler 2019, 7). In other words, the *kinds of* transgenerational, intergenerational, interpersonal, personal and even prepersonal[3] circuits that used to "emerge through affective relations of various kinds" over time within and across societies to forge "dreams, goals, objectives and common horizons" have been disrupted in *unprecedented* ways (ibid., 16). Rouvroy argues, in fact, that subjectivity is bypassed by contemporary digital automisation so that subjects are ren-

individuals or groups of individuals to the extent that the "reason of the state" becomes the reason/ing of the individual or groups of individuals (Foucault 1977, 68).

[3] Such as Deleuze's 'virtual' and Simondon's 'pre-individual field' or fund.

dered little more than a "collection of infraindividual data" that are "recomposed at a supra-individual level under the form of profile" (Rouvroy and Stiegler 2016, 12). Second, the speed at which what we have elsewhere termed the *Algocene*[4] has emerged, as well as the level of technological literacy a coherent understanding of this shift entails, has far exceeded our capacity to theorise its effects, with the result that many analyses of our contemporary condition remain trapped in anachronism, projecting the image of an old world and its struggles onto a new and largely alien terrain. It is perhaps not too hyperbolic to claim, as many have, that the staggering technological shifts of the last 20-odd years necessitate theoretical engagement on the scale of a new geological era or epoch. This is no longer the world of industrial capitalism, nor even of Foucault's disciplinary societies (Foucault 1977), Deleuze's control societies (Deleuze 1992) or Guattari's Integrated World Capitalism (Guattari 2000), although these are all vestigially inflected in the new situation, which has variously, although far from exhaustively, been termed societies of hyper-control and computational capitalism (Stiegler 2019), cognitive capitalism (Yann Moulier-Boutang 2012), platform capitalism (Srnicek 2016), the age of planetary computerisation (Guattari 2013), The Stack (Bratton 2015), infopolitics (Koopman 2018), instrumentarianism (Zuboff 2019) and algorithmic governmentality (Rouvroy and Berns 2013). Third, because of the speed at which digital innovation operates, whatever effects this new "hyper-synchronization of consciousness" (Vignola 2017, 188) is having on subjectivity can only be perceived in the most miniscule of ways, yet we have to rely on this minimum of information for a symptomatology because, if we follow Deleuze via Nietzsche, it is only through a symptomatology that we can trace etiological factors in order to find the most inventive corrective therapy. A difficult task indeed considering the constraints we are outlining here! The point is, we urgently require new tools to grapple with the implications of the encroachment of algorithmic reason and governmentality into more and more aspects of our lives and minds. Far from eschewing

[4] With reference to Deleuze and Guattari's *A Thousand Plateaus* (1987), Aragorn Eloff argues that the Algocene constitutes a new stratum, namely the algoplastic, and alters the functioning of order of words so as to enable novel modes of subjectivation. For more on this, see Eloff (2021).

theory in favour of the transcendent *mathesis universalis* of our new digital overlords, situated thinking about our times has never been more crucial. If, however, this kind of thinking has not been forthcoming to the extent we would hope for, then this too perhaps forms part of the symptomatology of the Algocene. In describing something that is unprecedented, we need to take care, as McKenzie Wark argues, to find "a renewed language for describing the present situation and identifying what in the received language of capitalism [or, for those further behind the times, Marxism] impedes forward movement in thought and action" (Wark 2019, 6). Having said this, we also hold that it is important not to reduce the entire social field to algorithmic control, despite its near-ubiquity. What is rather vital is to understand what this new way of experiencing subjectivity produces and what, in turn, produces it. As Guattari explains of the Freudian Unconscious, it *produced* new desires, which included "hysteria, infantile neurosis, psychosis, family conflict, the reading of myths, etc." (Guattari 1995, 10). So too what Stiegler refers to as "negative collective protention", or the nihilistic expectation of nothing, except perhaps The End (Stiegler 2019, 19, 50), has produced new desires, and here we have to seriously consider that The End might in fact—probably is—*desired*, even collectively so (especially given that we also face Anthropogenic doom).[5]

How, then, do we change this desire? How do we use the diagnostic tools at our disposal in the schizoanalytic mode—that is, a therapeutic mode—to produce something new, something healthier? How do we change our practices and ways of being in the world so that *life*—rather than a misguided idea of Algorithmic Supremacy—matters again? In this chapter, we attempt to give a broad symptomatology of algorithmic subjectivity, drawing on a range of scholars who have gone before us. Our

[5] Although we do not have the space here to discuss it, we have in other work looked at how the increase in excessive media and digital communications usage has created new forms of social inclusion and exclusion, of which the most exacerbated instantiation is probably the phenomenon known as *hikikomori*, first identified in Japan. Hikikomori refers to the acute, prolonged social withdrawal of individuals which has been linked to other societal variables, including the prevalence of precarious contract and part-time work, low skill sets and truancy. Hikikomori—a desire for *an* end—is part of our ongoing symptomatology to understand what we see as a general set of symptoms of society that we trace etiologically to algorithmic governmentality. For more, see Gray and Eloff (2021, forthcoming).

aim, diverging from most theorists, however, is not to propose legal or other reforms. We understand the algorithm to already present itself as an ethicopolitical arrangement of values, assumptions and propositions about the world. Our question—yes, our desire—is not: *How ought the algorithm be arranged for a good society?* but rather: *What is a good society?* For it is from the latter question, we hold, that healthier algorithmic arrangements will flow, not the other way around.

What Is It Like to Be a Generative Adversarial Network?

> *Behind the appearance of individuated subjectivity one must try to locate the real processes of subjectivation.* (Guattari and Rolnik 2007, 469)

It seems clear that new modes of subjectivation have emerged in the Algocene. Elsewhere, we have drawn parallels between these subjectivising processes and the form of computation/cognition exemplified by adaptive neural networks (ANNs)—a class of connectionist AI, and specifically the subclass of ANNs known as generative adversarial networks (GANs) (Eloff 2021). To give only the most non-technical synoptic gloss of the functioning of these systems, the ANN model operates by distributing a series of digital 'neurons' on a 'layer'. A deep convolutional network, the most common current form of this model, comprises multiple interconnected layers, including an input layer, where the data to be evaluated by the ANN is presented in a suitable format (typically an array of numbers/vectors), as well as an output layer, which represents the response of the network to the input data. Input data is initially used to 'train' the network by incrementally stabilising the weightings of neurons on each layer until a best fit or minimum error threshold has been reached and subsequently, in a trained network, to present new data for evaluation. Typically, the training process involves a fair amount of human intervention—selection of the training material, ongoing manual adjustments of weightings and thresholds and so forth—but GANs, or generative adversarial networks, largely bypass these forms of intervention by pitting two ANNs—a trainer and trainee—against each other. The trainer presents the trainee with inputs and attempts to 'trick' it, forcing it to learn from its mistakes, while the trainer itself simultaneously learns how to trick better, creating a high-speed iterative optimisation loop between the two ANNs (Goodfellow et al. 2014). This broad stroke account is not an entirely accurate representation of the technical details of convolutional neural networks (CNN),[6] but the key salient feature vis-à-vis subjectivation is that the GAN system reflects a mode of automatically self-optimising regulation/stabilisation behaviour that is echoed in what we observe as proto-autopoietic adaptations to algorithmic reason by human actors. In other words, just as the GAN contains a trainer and trainee, both of which are recursively modifying their behaviour relative to each other until a desired optimality is reached, so too can we view processes of subjectivation in the Algocene as consisting of reciprocal training. That is, the statistical aggregation of our data is modelled and presented back to us as an adaptive ideal—often with embedded social, existential and economic exigencies, for example access to work, healthcare and housing—and our subsequent modification of behaviour is then fed back to our digital trainer in a never-ending spiral that reduces our agency to a constant modulation of our data points while erasing those aspects of us that are not amenable to exhaustive quantification.

This "hyperindexation" causes human individuals to conceive of themselves as "hyperquantified" (Rouvroy and Stiegler 2016, 9)—what Deleuze calls the transition from individuals to "dividuals" (Deleuze 1992, 5). To put it differently, "algorithmic governmentality dividuates individuals by separating them from the cognitive, affective and empathic capabilities they possess and through which they can experience the world" (Vignola 2017, 191). Moreover, the transition to dividuation goes along with subjectivation processes of individual*isation* rather than individuation. Individualisation is thus a premature interruption—*disruption*—of individuation. This is a crucial distinction for Stiegler in his theorisation of the age of disruption. Stiegler's account of how individual

---

[6] A convolutional neural network (ConvNet/CNN) is a type of deep neural network or deep learning algorithm that most commonly analyses visual objects by assigning learnable weightings and biases to a variety of aspects of an input image and is also able to differentiate one (kind of) image from another.

subjects and objects come to be in the world, like Deleuze and Guattari's before him, is heavily influenced by the work of philosopher of science, Gilbert Simondon, particularly his highly influential doctoral thesis, now finally translated into English, almost 60 years later, as *Individuation in the Light of the Notions of Form and Information* (Simondon 2020). English speakers would have glimpsed some of his ideas in the 2017 translation of *Du mode d'existence des objets techniques* (*On the Mode of Existence of Technical Objects*), in which Simondon provides a general analysis of technicity[7] in its relation to human and nonhuman existence. In Simondon's novel account, which relies heavily on thermodynamics, quantum theory, cybernetics and various other cutting-edge scientific fields of his era, individual beings are understood as secondary to individuating dynamisms. What we understand as the everyday world of fully formed subjects and objects, argues Simondon, in fact conceals the primacy of flows and processes which, in turn, rely on an inexhaustible "pre-individual field", by which Simondon means, first, a spatial determination or topological fold "genetically prior" to the living individual or object (Simondon 1992, 306) and, second, a temporal determination that it is itself "the solution and dimension of the discovered systematic", rather than a "framework in which the genesis unfolds" (ibid., 315). In other words, the pre-individual field provides the "funds" for the unfolding of being.[8] This unfolding proceeds through a number of "phases", of which unity and identity of being are only one; it is only through grasping the full "dephasing" of being (Simondon 2009, 6) in these processes that we are able to arrive at a full ontogenetic[9] account that does not rely on untenable metaphysical assumptions. That is, an account wherein

[7] Technicity is key for Stiegler who understands "knowledge in general"—or what he also refers to as the "moral dimension of the noetic being" to be inscribed *not* "by God in the soul of this being", but rather as something that "must be learned and cultivated"—and it is precisely this dimension of learning and cultivation that he thinks of as the "technicity of life". Moreover, Stiegler views this dimension as "fundamentally accidental, artificial and contingent, stem[ming] from a primordial diversality … on the basis of which consistencies are projected … as protentions to the infinite … in the encounter with and counter to the incoherence of becoming" (Stiegler 2019, 227).

[8] If this all sounds rather familiar to Deleuzian ears, this is due to Simondon's enormous influence on the fourth and fifth chapters of *Difference and Repetition*—an influence that has only recently been fully acknowledged.

[9] Ontogenetic rather than ontological; that is, an accounting not only for being but for that which gives rise to being (the pre-individual field).

being is more rightly understood, to use Deleuze and Guattari's terminology, as becoming or heterogeneous multiplicity (Deleuze and Guattari 1987, 232). Instead of ontologically stable subjects and objects, individuation, given that the pre-individual funds (loosely equivalent to what Deleuze terms the "virtual") are not exhausted in any particular provisional phase of being, can be seen as a nonlinear metastability giving rise to systems that can always dephase into other holding patterns of equilibrium. As Elizabeth Grosz explains:

> In life, the processes of individuation never cease, they coexist with the duration of the living organism itself—the organism never fully coincides with itself, or attains an identity in which it is what it is. The living organism is more a singularity than an individual; and ironically, it is material individuals which attain the self-identity for which we assume a subject strives. (Grosz 2007)

In short, the individuation of an individual can be seen as a response to a *problematique* in a metastable system—so individuation provides a "partial and relative resolution" to this prior tension (Simondon 1992, 300). Individuation is also, if we follow Deleuze, a two-way flow: the emergent outcome of a process of becoming is influenced by the pre-individual, but so too is the network of relations that comprises a pre-individual field, which is the relation of difference to difference—a constant flux of self-differentiation—influenced by what is actualised from this flux in what Deleuze refers to as processes of counter-actualisation (Deleuze 1990, 150).

Simondon analyses processes of individuation across multiple scales of organisational complexity, beginning with simple non-organic physical systems and then proceeding to simple organic systems before exploring the psychic individuation of minds and the collective individuation of forms of social organisation. While the specifics vary based on the system under discussion,[10] in all instances, these processes can be viewed as pro-

[10] For example, organic systems internalise various parts of their milieu or processes of becoming, whereas non-organic systems rely more strictly on external processes located at their outer limits—we may think here of crystallisation, which occurs based on differences between the outer edges of a seed crystal and the supersaturated solution that surrounds it.

viding a partial solution to the tension of various disparate levels—a resolution of disequilibrium or *disparation*, or what Simondon describes as *transduction*, which results in the production of *information*. Transduction, in Simondon's understanding, is an order of communication that liberates and actualises the potential energy in a pre-individual system (see Simondon 2017, 156). As such, it provides the structure for the eventual actualisation while also contingently conditioning that actualisation or ontogenesis of an individual—be that individual physical, psychic, social or technical (Simondon 2009, 11). Crucially, for our current purposes, transduction gives rise not only to the individuated being but, simultaneously, to the milieu within which individuation takes place, just as for Deleuze and Guattari becomings are always reciprocal (Deleuze and Guattari 1987, 239). This is especially critical for Stiegler's theorisation of transindividuation and the relationship between the individual and collective retentions of 'noetic' beings,[11] along with their reliance on the tertiary retention of technical objects and how this relationship is disrupted by algorithmic governmentality.

With reference to Horkheimer and Adorno's seminal work on the culture industry, Stiegler argues that "the processes of psychic and collective individuation characteristic of the life of the mind and spirit have slowly but surely been wiped" and are "now exclusively operating in the service of the market and the organization of consumption" (Stiegler 2019, 5). This disruptive process "penetrates, invades, parasitizes and ultimately destroys social relations at lightning speed, and, in so doing, neutralizes and annihilates them from within, by outstripping, overtaking and engulfing them" (ibid., 7). For Stiegler, then, *disruption* is the

[11] "We are noetic beings to the extent that we weave psychic secondary retentions on the framework of collective secondary retentions, constituted from psychic and collective preindividual funds: we individuate ourselves by exteriorizing the protentions contained within these retentional funds, hidden as 'potentials' that are 'concretized' and 'actualized' through being transindividuated" (Stiegler 2018, 34). Also: "When noetic individuals live the time of an experience in which they select traits as primary retentions on the basis of secondary retentions, they at the same time and in return interpret these secondary retentions insofar as they form ensembles. These ensembles are charged with protentions derived from previous experiences. Some of these protentions are transindividuated and transformed into a common rule, that is, into habits and conventions of all kinds, metastabilized between the psychic individuals and the collective individuals associated with these experiences (a con-vention being what con-venes a plurality of individuals: it is what gathers their coming together)" (Stiegler 2016, 140–141).

"destruction of social systems by the technical systems" (Stiegler 2019, 37); the dramatic short-circuiting of the processes (or "long-circuits"[12]) of psychic and collective individuation that have historically given rise to subjectivities. In simpler terms, algorithmic governmentality, unleashed by our contemporary technical systems and the technocrats—or "new barbarians"—in charge, are decimating social systems and 'hacking' the tertiary memory of psychic individuals through an unprecedented "proletarianization" of the mind (ibid., 11, 8). These processes of proletarianisation, which alienate individuals not just from labour but from individuation itself, deprive humans of knowledge, and especially of long-term and intergenerational knowledge: "in the sense of reason understood as logical", but also "of life-knowledge [*savoir-vivre*]" and "work-knowledge [*savoir-faire*, knowledge of how to make and do]"—our very ways of existing and being in the world (ibid., 179). The result is an unparalleled crisis of meaning and the subjection of the future to "an absolutely negative protention" (ibid., 10) that Stiegler refers to as the "absence of epoch" (ibid., 10–18). It is nothing less than Deleuze's missing people negatively deterritorialising into Stiegler's missing epoch. If these critical concerns around technology initially cause the reader to view Stiegler in a reactionary light, we should remind ourselves that in his work—as in Simondon's—technics has always been a part of the coming-into-being of humanity. To clarify his view in this regard, we need to take a brief detour through his engagement with Husserlian phenomenology and his understanding of *retention* and *protention*.

For Husserl, briefly, we gain our awareness of time, or our *time-consciousness*, through what he calls a triple intentionality: beyond the present moment (the *primal impression*) there is both the immediate past of *retention* and the anticipated future of *protention*. These instances, however, are not experienced as discrete; rather, they are bound together in a single duration. That is, the most granular level of any actual phenomenological experience contains all three components of the 'present' (Husserl 1991, 21–28). As Merleau-Ponty describes it: "The emergence of a present now does not provoke a piling behind of a past, and a pulling of the future. Present now is the slippage of a future to the present, and

---

[12] See Stiegler and Rogoff (2010).

the just-past to the past: it is in one single movement that time sets in motion in its entirety" (Merleau-Ponty 2002, 479). The example of a musical melody is famously used (Husserl 1991, 22–25) to explain this account: when we hear an individual note, our primal impression of that specific sonic frequency is contextualised by what went immediately before it, as well as the possible notes we expect to immediately follow it. If the pianist plays an E, we will hear this E differently based on whether the previous note was a C or an A, and this process is incremental, with each subsequent note pushing the previous ones back along the chain of retentive influence upon our immediate impression.[13] The question, though, is why we anticipate a particular note or set of notes to follow those we are currently hearing. If our immediate experience of the passing flow of time was all there was, we would have no ability to anticipate anything at all or even, as Stiegler argues, to "select"[14] which aspects of the primal impression to pass into retention. Stiegler, following Husserl, understands primary retention as "the material of perception, and therefore of the present inasmuch as it presents itself, which is to say that the present is a dynamic process of presentation", and secondary retention as "memory, that is, of the past, of what is absent and represented by a dynamic process of imagination" (Stiegler 2019, 215). Following from this, Stiegler understands the criteria for primary retention and protention to be constituted by secondary retention (or memory, loosely), which is itself formed through the accumulation of past experience. Hence, we anticipate a G to come after the E because we have heard a particular song or similar progressions before. Stiegler expands on this interplay[15] by reference to Simondon's work on psychic and collective individuation, and his account of collective secondary retentions and protentions is

[13] It is worth noting that this account of consciousness of the present enjoys some neurobiological support, particularly in accounts like those of Francisco Varela, within whose neurophenomenological approach triple intentionality finds its biological analogue in neuronal ensemble synchronisation (Varela 1999). Incidentally, the underlying neuronal model has also been applied to some artificial neural network models.

[14] See Stiegler (2019, 216).

[15] Stiegler modifies both Husserl's account and Derrida's critique of the separation of primary and secondary retention in various ways, but a full elucidation of this is beyond the scope of this chapter.

crucial for his development of the idea of tertiary retention, a core element of his theorisation of technics. Stiegler clarifies as follows:

> When noetic individuals live the time of an experience in which they select traits as primary retentions on the basis of secondary retentions, they at the same time and in return interpret these secondary retentions insofar as they form ensembles. These ensembles are charged with protentions derived from previous experiences. Some of these protentions are transindividuated and transformed into a common rule, that is, into habits and conventions of all kinds, metastabilized between the psychic individuals and the collective individuals associated with these experiences [while others] continue to await transindividuation, that is, expressions and inscriptions that continue the development of already-existing circuits of transindividuation. (Stiegler 2016, 140–141)

To summarise the argument so far, Stiegler holds that the binding of time by primary forms of sense perception and anticipation is bolstered by secondary retentions and protentions of memory and imagination that are both individually held as well as collectively maintained *across time* by social groupings, such as families and cultures. Cross-generational institutions, traditions, practices and so on are thus seen as a form of shared memory. Crucially, Stiegler, who also refers to the primary as *epigenetic* and the secondary as *phylogenetic*, fundamentally modifies the Husserlian account at this point with the work of paleoanthropologist Andre Leroi-Gourhan, observing that we have, at least since the Upper Palaeolithic, spatialised and materialised our accreted retentions or shared memories by embedding and exteriorising them in *epiphylogenetic* technological artefacts like tools, manuscripts, buildings and so on (Stiegler 1998, 135). Given that the *memory traces* of technological artefacts, or *hypomnemata*, provide "the force of synchronization and the diachronization of memory" as a form of "compensation to the retentional finitude of the organism, since an organism cannot retain all its experience and cannot transfer this experience to the next generation without having exteriorized them as symbols and tools" (Hui 2019, 202), Stiegler insists that we need to include technics as a form of tertiary retention. In fact, tertiary retention is a primary part of subjectivation because it constitutes technologically embedded memory that sustains both living cultures and the individuals that move in and through them. Moreover, processes of *epiphylogenesis* are what allow us to experience ourselves as time-binding subjects, to "form a circuit in which the soul is no longer simply a movement that returns to itself to determine itself … but also a *tekhnesis*, whose organization depends on the third memory" (ibid., 201). To return to our musical example, it is the tertiary retention of the CD or mp3 that invokes primary and secondary retentions and protentions whereby the individual binds the present moment into *meaningful experience*. Or, as Stiegler argues, "tertiary retention always already precedes the constitution of primary and secondary retention. A newborn child arrives into a world in which tertiary retention both precedes and awaits it, and which, precisely, constitutes this world as world" (Stiegler 2009, 9).[16]

Stieglerian interlocutor, Yuk Hui, further expands upon the Husserlian-Stieglerian schema to include *tertiary protention*, which is the manner in which technics not only retains the past but also (contingently) structures the ways in which we can anticipate the future (Hui 2019, 38). In the context of the Algocene, recursive algorithmic processes give rise to a unique form of digital tertiary protention which "cannot be explained merely in terms of the anticipations of a subject, but rather requires analysis of the forms of anticipation made possible by algorithms, for example, different systems of recommendation or applications of machine learning" (Hui 2018, 147). In this situation, "tertiary protention becomes increasingly active, to the point of displacing or marginalizing active directedness" and these protentions are "characterized by their directedness to the future" (ibid.). To elucidate the notion of digital tertiary protention, Hui provides the useful example of an online restaurant booking:

> When people want to go to a restaurant, these days they are increasingly likely to search online first. We might also notice that Google is able to suggest which is the closest and most preferable restaurant for their needs according to its search and recommendation algorithm. We can make at least two primary observations based on this example: (1) tertiary

[16] The similarities between Stiegler's account of tertiary retention and the extended mind hypothesis developed by Andy Clark and David Chalmers, which forms part of the popular 4E account of cognition, should not be overlooked (Clark and Chalmers 1998).

> protention tends to depend on tertiary retention, for example, the relations given by digital objects, those traces we have left, such as pictures, videos, or geolocations; and (2) orientation becomes more and more an algorithmic process that analyzes and produces relations to pave the way for the experience of the next now or the immediate future. (Hui 2016, 221)

Examining the implications of this digital dynamic, Hui goes on to argue that the widespread nature of Big Data analysis and predictive computing via contemporary AI, as well as the normativity this gives rise to through what we have been terming algorithmic reason, eradicates—or at least significantly marginalises—human agency from the protentional cycle with the result that, as Stiegler concurs, both cultures and individuals submit the future to the *mnemotechnologies* (Stiegler's term for contemporary forms of tertiary retention) and modes of technological rationalisation of the eternally self-optimising autopoietic present. As such, there is a fundamental collapse of possibility and meaning onto the homogenising, collectively and psychically dis-individuating horizon of the endless now. In losing our capacity to form protentions, we simultaneously lose our knowledge of how to live and it is our contention that this symptom can be traced etiologically, at least in part, to the problem of social inscription or *epigraphy*, which is closely related to Stiegler's understanding of processes of epiphylogenesis and how these allow us to experience ourselves as time-binding subjects.

Early forms of social inscription, or what Deleuze and Guattari refer to as "territorial" inscriptions, covered a social field by means of marking—be it physical markings like tattoos, brandings and scarifications, or the wearing of masks and the performance of rituals. This kind of inscription can be thought of as "the collective investment of the organs" because "flows are coded only to the extent that the organs capable respectively of producing and breaking them are themselves encircled, instituted as partial objects, distributed on the socius and attached to it" (Deleuze and Guattari 1983, 142). This necessarily allows for 'group fantasies' or societal dreams and objectives because there is a *collective investment of desire*. More importantly, this collective investment of desire and its means of inscription produces a form of collective memory vis-à-vis the inscription practices of a society or culture. Socio-cultural memory thus becomes a site of particularised yet heterogeneous long-circuits of transindividuation, allowing for "the passage of thought across time", a connection with even very "distant generations" that creates an accord with the past, in turn "creating a unity with the future" (Stiegler and Rogoff 2010, 1, 2). Although the primacy of inscription is temporal, as we see here, it is a temporality that exceeds the life of an individual or even a group of individuals (generation)—and this is the most salient aspect for Stiegler because it involves the technics of tertiary retention. (It is helpful here to even think of inscription as a technics of spatialised transmission.) Deleuze and Guattari warn, however, that "the inscription on the socius is in fact the agent of a secondary psychic repression" that is "situated in relation to the desiring-inscription of the body without organs, and in relation to the primary repression that the latter already performs in the domain of desire—a relation that is essentially variable" (Deleuze and Guattari 1983, 184). Basically, any inscription concurrently produces a tertiary retention and an apparatus of repression—or we could say, it functions as a *pharmakon*—but the latter can vary greatly. Primitive or territorial inscription, for example, retains quite a high degree of polyvocity but produces *debt* as the apparatus of repression (ibid., 184–185). The despotic period[17] then sees a greater convergence and homogenisation of flows of desire, funnelled by the "'megamachine' of the State" which overcodes "the old inscription" (ibid., 194, 196), extending and reinforcing older regimes of debt. The colonial/imperial era too has its own distinct signs. On the one hand, the law—a "paranoiac-schizoid" metonymic apparatus—comes more fully into being to administer, that is *bureaucratise*, the State apparatus. What is more, Deleuze and Guattari tell us, is that the law, besides expressing a paranoiac-schizoid aspect, also articulates a "maniacal depressive trait (metaphor) according to which" it "reveals nothing and has no knowable object, the verdict having no existence prior to the penalty, and the statement of the law having no existence prior to the verdict" (ibid., 212). Hence debt, which became infinite debt, now assumes a juridical form and *desire* becomes more deeply entrenched in the State—what might be thought of as the inscription of

[17] This 'era' should not be viewed as discreet but, rather, as intersecting and overlapping with earlier and later 'periods', even though novel alliances and filiations can be found to 'mark' it.

Nietzsche's *ressentiment* and, one step later, of Oedipus: "the biunivocalization, application, and linearization of the chain between masters and slaves; the introduction of the law into desire, and of desire into the law; the terrible latency with its afterward or its after-the-event" (ibid., 215). The capitalist machine, far from overcoding, *decodes* flows, processes and desires, discharging a breath of freedom, only to axiomatise the moment of liberty. We know this part well: in this "system no one escapes participation in the activity of antiproduction that drives the entire productive system" (ibid., 236). We are all slaves to capital, but even so, there is a thread, a story, a trajectory, revolt even and a collective memory. In our contemporary societies, however, we face a completely novel situation where the deployment of algorithmic processes and infrastructures across the social and psychic landscapes short-circuits our psychic and collective protentions (long-circuits of transindividuation, or collective memory) and replaces them with algorithmically generated protentions—or what Stiegler calls "digital tertiary retention" (Stiegler 2018, 96)—which form the training component of the generative adversarial networks our individual and social lives complement as that to be trained.[18] This process is hard to resist given the incredible speed at which contemporary computing and communication technologies function. How can our psyches resist being "outstripped, overtaken and progressively replaced by automatic protentions" when these operate at "between one and four million times quicker than the nervous systems of psychic individuals"? (Stiegler 2016, 131).

One consequence of this hyper-synchronisation of individual and collective life and digital tertiary retentions and protentions is what Stiegler calls *hyper-diachronisation*—the dispossession of our capacity for "desires, expectations, volitions, will and so on" (Stiegler 2019, 7), and thus a collapse of any sense of self or group agency into the passive provision of data, powerfully captured in his reflection on many teenagers who no longer dream. The outlook is bleak: shorn of our sense of meaningful participation in the world as individuals, communities and so on, new forms of thinking and praxis are unable to emerge and we are instead

---

[18] In case this seems opaque, consider one of the simplest examples of digital tertiary protention: autocomplete.

bound to simply repeat the present. To put it in Deleuze's terms, the possibilities for the kind of 'shock to thought'[19] that is needed to rouse us from 'stupidity' are fast disappearing, leading to a situation where, Stiegler warns us, transindividuation will become impossible and we will become lost in abject aboulia. The problem, as we have described it in relation to inscription, is that inscription needs a surface—be that the marked body, the marked psyche or the marked society (rituals and other externalised technics of transmission)—which locates an affective decussation, a "distinct mark that has a place within a general system of marks", together with a "difference in intensity that is registered in the unconscious" as Jon Roffe says in a somewhat different yet related context (Roffe 2015, 54). These affective memory-traces or intensities are at once individual *and* collective because they cut or decussate a history, *entire histories* sometimes. The cloud, with its vastly diffused composition and the speed at which it operates, leaves hardly any surface for tertiary retention, and so erases our technics of life; that is, the current apparatus of repression supresses psychic and collective protentions by replacing them with digital tertiary protentions. With the collapse of relations of co-individuation and social processes of transindividuation, and the concomitant erasure of inscription surfaces, there is no basis for sustainable or intergenerational social systems and we thus lose the ability to produce shared meaning and sense, desires and hope.

It is worth reminding ourselves here of the close imbrication of digital tertiary retention/protention and contemporary capitalism. For Stiegler, the algorithmic capture of our time-binding "constitutes the epistēmē[20] of capital", which is in fact an "anti-epistēmē because, as instrument of hyper-control and generalised proletarianisation, it amounts to the most advanced stage of capitalism qua process of proletarianization; that is, as a process that destroys knowledge" (Stiegler 2018, 147). This draws us close to Guattari's analysis, also influenced by Simondon, of processes of desingularisation under contemporary capitalism. Indeed, if we apply Guattari's critique of the homogenisation of valourisation to the Algocene

[19] Deleuze discusses this in *Cinema 2: The Time-Image* (Deleuze 1989, 156).

[20] The anti-episteme is also the loss of epoch if, like Stiegler, we see an epoch as "what enables collective protentions to be established through the constitution of new circuits of transindividuation" (Stiegler 2019, 15).

and place it alongside Stiegler's symptomatology,[21] we can see how the contemporary digital "capitalist system of valorization, in which value is based upon a general equivalent" (Guattari 1989, 146) and in which "all other modes of valorization, which thus find themselves alienated from capitalist hegemony" (ibid.) are crushed, is an application of the same hubris of *mathesis universalis* that gives rise to the generative adversarial network model of (inter-)subject-ivation. The digital automation of the 'reason' of the market and its drive to reduce everything to quantification find a close counterpart in the automation of retention and protention that are increasingly dissolving social and subjective bonds into a morass of data points and optimisation problems, and producing "a subjectivity which seems likely to blot out, with its greyness, the faintest traces and last recesses of the planet's mysteries" (ibid.). All that is solid melts into air, the air thick with data streams, hardly a surface in sight. Armed with this partial and provisional symptomatology of the present, let us now turn to an examination of its manifestations in social media.

The Subjugated (Disindividuated) Groups of Social Media

> *[T]echnologies of capitalism, from the assembly line to Facebook, have created processes of 'disindividuation,' short-circuiting the productive becoming of individuation while resulting in desublimation. Thus, the infinite potential of becoming is continuously compressed into the finite gestures of capitalist proletarianization.* (Galloway and Larivière 2017)

The various forms of online life are perhaps the quintessential example of the contemporary processes of desingularisation we have been discussing thus far; indeed we will argue that the very fundamentals of subjectivity, intersubjectivity, agency, ethics, politics and ontology, to name a few, are being transformed by the near-ubiquity of various forms of social media and digital interaction. To begin, we need to remind ourselves that

[21] Or, as Guattari says, make sure to situate "the concrete incidence of capitalistic subjectivity (the subjectivity of generalised equivalence) within the context of the continued development of the mass media, Collective Equipment and the information revolution" (Guattari 1995, 22).

social media has, to an unprecedented extent, led to what sociologists refer to as "context collapse".[22] Whereas previously our social lives played out in multiple overlapping but heterogeneous contexts (family, friends, school, work, etc.) and we cultivated and exhibited different aspects of our identities in each of these contexts,[23] today all of these contexts are drawn together by massive social media platforms with billions of members. Facebook, Twitter, Instagram and similar platforms do not distinguish between work colleagues, family members and ex-partners, and our participation in these highly overexposed spaces is always punctuated with a subtle sense of discomfort that the entire global agora is listening in. Of course, over time the extent and composition of this crowd are to some degree determined by our patterns of online behaviour. That is, the various search engines and social media platforms use algorithms to incrementally adapt what they present to us as an impartial view on reality based on how we interact with those platforms and others on them, with the result that many social networks often end up starkly homophile—an increasingly commonplace kind of ideologically and politically homogeneous 'echo chamber' that can breed noxious ideologies, of which the recent QAnon[24] phenomenon is a good example. Not only does this algorithmically modulated incrementation adaptation of our online realities then go on to inflect our real-world lives—which purchases we make, which protests we attend, who we vote for—but it also, through the aggregation of our data and its re-presentation back to us in a constant iterative feedback loop, nurtures the digital protentional cycle we discussed above, alienating us from our sense of self and agency, which we are then encouraged to substitute through Manichean forms of online group belonging that lack much of the nuance and uncertainty of real-

[22] See, for example: Davis, Jenny L. and Jurgenson, Nathan. 2014. "Context Collapse: Theorizing Context Collusions and Collisions". *Information, Communication & Society* 17(4): 476–485, DOI: 10.1080/1369118X.2014.888458.

[23] Not that these contexts did not have their own problems—just think of Foucault's disciplinary societies. The point is rather that despite the problems, there was a kind of continuity that has been eroded.

[24] Preceded by the 'Pizzagate' conspiracy theory and overlapping with it in some ways, QAnon is a far-right conspiracy theory alleging that a group of demonic, paedophilic liberal politicians running a sex-trafficking ring rules the world and is plotting against Donald Trump, astonishingly cast in a messianic role against the 'cabal'.

world equivalents. Indeed, the performance of identity online tends to be highly unequivocal, sectarian and hyperbolic so that we become increasingly like the simplistic avatars meant to represent us in digital form, identifying absolutely with overwrought social, political and existential binaries which, in turn, inflects the identities we carry across to our offline interactions. The tone and content of much of the discourse emerging from the recent Trump presidency, as anyone who has watched video interviews with crowds of Trump supporters can attest to, exemplify this becoming-avatar-of-subjectivity. Similarly, the encroachment of some of the more harmful elements[25] of 'cancel culture' into the real world reflects an internalisation of the norms and codes of conduct in online spaces—such as appeals to admins, banning or blocking of those we no longer wish to hear from, asserting belonging or non-belonging in totalising either/or fashion—and their enactment in contexts where they are ill-situated.

We can also observe a certain kind of 'emoticonification' of affect that has resulted from online subjectivation. Think here of what happens when crucial processes of individuation unfold in contexts where affective states can only be reliably expressed using simple—and preselected—visual depictions of highly folk-psychological states, like, 'sad face', 'happy face', 'cross face'. These default online states are then transferred to real-world interactions where the tendency becomes to reduce nuanced affective positions to internalised emoticons. The result is that so-called extremely online people, who we can also, to a large extent, refer to generationally, struggle with expressing complex emotions and with delayed gratification, especially when the short-term neurochemical (dopamine) 'hits' associated with the expression of these simplified emotional responses, along with simplified forms of social belonging and anti-social phenomena like 'shitposting'—a powerful example of what Stiegler calls *je-m'en-foutisme*, or 'I-don't-give-a-fuckism' (Stiegler 2019, 98)—have to a large extent been employed as psychological palliatives to cover up the sense of anxiety, dread and pointlessness stemming from the loss of

[25] To underscore, there is of course merit in approaches that seek to redistribute power and agency by amplifying previously marginalised voices and removing platforms from voices that seek to perpetuate imbalances in power and agency; here we are certainly not aligning with a certain kind of right-wing reactionary sentiment that rages against a strawman of 'identity politics gone mad'.

meaning and direction in life. Broadly speaking, our contemporary collective relations to meaning, truth, authority, identity, affect and so on are increasingly reflective of the dynamics and limitations of interaction of online life, not just for the simple reason that we are spending more time online but, to return to Stiegler, because online life, as an immense high-speed generative adversarial network of retentional and protentional automation, is a primary locus of desingularisation and the erasure of collective memory. Truly, as Guattari would say, we are dealing with the subjugated groups of social media, and reflecting on this is becoming increasingly difficult in a context of infinite speeds where there is only time to adjust or adapt to the present, which societal transindividuation, ultimately leading to a situation where "the possibility of posing questions" is gravely hindered (Stiegler 2013, 105). That this universally ungrounding dynamic, wherein our pre-individual funds have been exhausted and replaced with algorithmic optimisation, has been correlated to the global growth of the right wing[26] is no surprise when we consider that, historically, right-wing discourse presents itself precisely via a discourse of certainty and grounding, Manicheanism and absolute categorial belonging. Relatedly, the very form of our thought has been altered, which in turn exacerbates factionalism.

None of this is separate from the alienating and dis-individuating dynamics of contemporary capitalism. As Guattari reminds us, whatever "the appearances of free thinking the new capitalist monotheism liked to cloak itself in, it has always presupposed an irrational and archaicizing hold over unconscious subjectivity", especially vis-à-vis "apparatuses of hyperindividuated responsibilization and culpabilization which, pushed to their paroxysm, lead to self-harming compulsions and morbid cults of blame" (Guattari 2013, 11). Together, the socially corrosive effects of the Algocene and contemporary capitalism collapse our psychic selves, our material lives, our communities and our collective and individual hopes and dreams, loves and desires, memories and reflections, into a vast self-modulating void of being, bound to perpetually repeat the present

[26] Beyond a shift to the right and the attendant shift of the Overton Window—or the range of acceptable political policies and social discourses—we call also see the results in phenomena like *hikikomori*, or extreme social withdrawal. For more, see Gray and Eloff (2021, forthcoming).

without any recourse to the pre-individual funds or societal inscription, now overdetermined by planetary digital networks of hyper-control, and in this void, all of the ills of the past emerge once more. With this perilous prediction in hand, how do we fabulate the future?

Fabulation: Long-Circuiting the Algocene

> *The same goes for Bartleby: the rule would lie in this logic of negative preference, a negativism beyond all negation.* (Deleuze 1997, 71)

The answer to our problems does *not* lie in primitivism or any other return to a (mythical) better past. We are here; we need to find ways to deal with *now* if we are to rescue the future. The answer, equally, does not solely lie in reform. Rather, we need to ask what we want our societies to look like—and here we can look to the past for some ideas without returning to it. In short, we need to foster new forms of collectivity in the maze of individualism and individualisation. This requires radical new formulations of subjectivity that take into account our contemporary situation. Like Bartleby, it might consist of a *preference not to*, which is neither refusal nor acceptance: it simply "posits an impossibility" (Deleuze 1997, 71). This impossibility or nonpreference produces a tension, an ambiguity that creates a path for escaping the neurosis of the world. That is, the little fissure it creates gives way to a mapping of trajectories away from subjugated groups and towards new types of group subjects that do not merely concede to the present, either in terms of an eschatology of techno-singularitarianism or hikimomori-style withdrawal. For Guattari, this involves processes of resingularisation, where singularisation refers to

> a self-organizing process that at its most basic level concerns bringing together ensembles of diverse components (material/semiotic; individual/collective), that is, assemblages … that deploy their own intrinsic references (inventing relations with the outside as well), and the analysis of their effects (especially transformations) on the formation of subjectivity beyond

> the individuated subject and prefabricated versions of him/her. (Genosko 2001, 129)

Far from the homophile groups of online life, this requires, as Guattari foresaw, an injection of heterogeneity where individuals can be confronted with ideas dissimilar from their own and opportunities to talk this through rather than reactively cancelling[27] someone. It requires forms of offline life that nurture transgenerational, intergenerational, interpersonal, personal and pre-personal long-circuits in order to reach the Neganthropocene, a place of dreams. As Stiegler puts it:

> we must turn to the question of the dream in politics and the question of the politics of dreaming—where dreams would be the resource of any neganthropogenesis, that is, any exosomatic organogenesis such that it would preserve its future by maintaining its noetic capacity. (Stiegler 2019, 162)

These new social practices of liberation—yes, of dreaming—cannot be based on new hierarchical relations. Instead, one of the main goals needs to be the creation of new alliances, a transversal politics, "that confers to these initiatives a character of living subjectivation and irreplaceable experience, that 'is worth being lived', that 'gives meaning to life'" (Guattari and Negri 2010, 125). In short, we are looking for health. Interestingly, for Deleuze and Guattari, as for Nietzsche before them, health is not opposed to illness; rather it is the "immanent process whereby illness brings about its own autocritique and self-overcoming" (Tynan 2010, 155). The illness of our time therefore needs to be transmuted in order to create new ways of living again—and this is of course the schizoanalytic practice. The aim is to see our current algorithmic articulations as a *pharmakon*—that, precisely, is Bartleby's stance. It is a

[27] To reiterate, we *do* think that cancelling does have merit when it seeks to redistribute power and agency by amplifying previously marginalised voices and removing platforms from voices that seek to perpetuate imbalances in power and agency. We are referring here to the overuse of cancelling as a way to avoid confronting views that may be dissimilar from our own—which is *healthy* to be confronted with.

productive paradox, an affirmation of life and the future, rather than a negation of it. Importantly, though, it is

> only *with* the pharmakon that knowledge takes *care* of the *pharmakon* … *the whole manifold of life-knowledge*—which passes through rituals as well as habitus, trades and forms of worship [*cultes*] that are found in any society insofar as it consists in a social cohesion that is a collective individuation, cultivating, through these always diverse pathways, the organic and organological solidarity in which it consists. (Stiegler 2019, 224)

The individualist, individualising societies we currently find ourselves in have not only lost their means of tertiary protention, but a memory of what a society is and can be: that which allows the development and co-creation of individual and social freedom; where material goods are produced in order to meet the physical needs of everyone in a society, rather than satisfy the insatiable requirements of capitalism; where free and equal social relationships are fostered so that people can flourish mentally and emotionally and, in turn, develop their ethical and creative capacities.[28] If this is the basis of our society, algorithms would have a very different function and *that*, we hold, is more important than reform.

References

Amoore, Louise. 2020. *Cloud Ethics: Algorithms and the Attributes of Ourselves and Others*. Durham and London: Duke University Press.

Anderson, Chris. 2008. The End of Theory: The Data Deluge Makes the Scientific Method Obsolete. *Wired*, 23 June. Available online: https://www.wired.com/2008/06/pb-theory/.

Barbrook, Richard, and Andy Cameron. 1996. The Californian Ideology. *Science as Culture* 6 (1): 44–72. https://doi.org/10.1080/09505439609526455.

Bogue, Ronald. 2010. *Deleuzian Fabulation and the Scars of History*. Edinburgh: Edinburgh University Press.

Bratton, Benjamin H. 2015. *The Stack. On Software and Sovereignty*. Cambridge and London: MIT Press.

---

[28] Inspired by https://voidnetwork.gr/2018/02/12/anarchist-society-look-like-anarchist-faq/.

Clark, Andy, and David Chalmers. 1998. The Extended Mind. *Analysis* 58 (1): 7–19. https://doi.org/10.1093/analys/58.1.7.

Couturat, Louis. 1901. *La Logique de Leibniz*. Paris: Felix Alcan.

Deleuze, Gilles. [1967]1991. *Masochism: Coldness and Cruelty*. Translated by Jean McNeil. New York: Zone Books.

———. 1987. *A Thousand Plateaus: Capitalism and Schizophrenia*. Translated by Brian Massumi. London and Minneapolis: University of Minnesota Press.

———. 1989. *Cinema 2. The Time-Image*. Translated by Hugh Tomlinson and Robert Galeta. London and Minneapolis: University of Minnesota Press.

———. 1990. *The Logic of Sense*. Translated by Constantin Boundas. London: The Athlone Press.

———. 1992. Postscript on the Societies of Control. *October* 59: 3–7.

———.1997. Bartleby; or, The Formula. In *Essays Critical and Clinical*. Translated by Daniel W. Smith and Michael A. Greco. Minneapolis: University of Minnesota Press.

Deleuze, Gilles, and Félix Guattari. 1983. *Anti-Oedipus: Capitalism and Schizophrenia*. Translated by Robert Hurley, Mark Seem, and Helen R. Lane. London and Minneapolis: University of Minnesota Press.

Eloff, Aragorn. 2021. 2006: The Topology of Morals (Who Does the Algorithm Think We Are?). *Deleuze and Guattari Studies* 15 (2): 178–196.

Foucault, Michel. 1977. *Discipline and Punish. The Birth of the Prison*. Translated by Alan Sheridan. London: Penguin Books.

Galloway, Alexander R., and Jason R. Larivière. 2017. Compression in Philosophy. *Boundary 2* 44 (1): 125–147. https://doi.org/10.1215/01903659-3725905.

Genosko, Gary. 2001. An Introduction to 'Singularization and Style: Shin Takamatsu in Conversation with Félix Guattari'. *Parallax* 7 (4): 128–130. https://doi.org/10.1080/13534640110089465.

Goodfellow, Ian, Jean Pouget-Abadie, Mehdi Mirzau, Bing Xu, David Warde-Farley, Sherjil Ozair, Aaron Courville, and Yoshua Bengio. 2014. *Generative Adversarial Nets*. In Proceedings of NIPS, 2672–2680. papers.nips.cc/paper/5423-generativeadversarial-nets.pdf.

Gray, Chantelle, and Aragorn Eloff. 2021 (forthcoming). The Fourth Ecology: Hikikomori, Depressive Hedonia and Algorithmic Ubiquity. In *Studies in Online Civilization, Volume 2*, ed. by Mirlea Sacks et al. Baltimore, Maryland: Montagu House.

Grosz, Elizabeth. 2007. Deleuze, Bergson and the Concept of Life. *Revue internationale de philosophie* 241 (3): 287–300. https://doi.org/10.3917/rip.241.0287.

Guattari, Félix. 1989. The Three Ecologies. *New Formations* 8: 131–147.
———. 1995. *Chaosmosis: An Ethico-aesthetic Paradigm*. Translated by Paul Bains and Julian Pefanis. Bloomington and Indianapolis: Indiana University Press.
———. 2000. *The Three Ecologies*. Translated by Ian Pindar and Paul Sutton. London: The Athlone Press.
———. 2013. *Schizoanalytic Cartographies*. Translated by Andrew Goffey. London: Bloomsbury Academic.
Guattari, Félix, and Antonio Negri. 2010. *New Lines of Alliance, New Spaces of Liberty*. Translated by Michael Ryan, Jared Becker, Arianna Bove, and Noe Le Blanc. New York: Minor Compositions.
Guattari, Félix, and Suely Rolnik. 2007. *Molecular Revolution in Brazil*. Translated by Karel Clapshow and Brian Holmes. New York: Semiotext(e).
Hui, Yuk. 2016. *On the Existence of Digital Objects*. London and Minneapolis: University of Minnesota Press.
———. 2018. Archives of the Future—Remarks on the Concept of Tertiary Protention, Inscription. In *Inscription*, 129–151. Gutenberg: Landsarkivet i Göteborg.
———. 2019. *Recursivity and Contingency*. Maryland: Rowman & Littlefield Publishers.
Husserl, Edmund. 1991. On the Phenomenology of the Consciousness of Internal Time (1893–1917). In *Collected Works, vol. 4*. Translated by John Brough. Dordrecht: Kluwer Academic Publishers.
Koopman, Colin. 2018. Infopolitics, Biopolitics, Anatomopolitics: Toward a Genealogy of the Power of Data. *Graduate Faculty Philosophy Journal* 39 (1): 103–128. https://doi.org/10.5840/gfpj20183914.
Mazzocchi, Fulvio. 2015. Could Big Data Be the End of Theory in Science? A Few Remarks on the Epistemology of Data-Driven Science. *Embo Reports* 23 (10): 1250–1255. https://doi.org/10.15252/embr.201541001.
McQuillan, Dan. 2018. Data Science as Machinic Neoplatonism. *Philosophy & Technology* 31: 253–272.
Merleau-Ponty, Maurice. 2002. *Phenomenology of Perception*. Translated by Donald A. Landes. London and New York: Routledge.
Moulier-Boutang, Yann. 2012. *Cognitive Capitalism*. Cambridge: Polity Press.
Pigliucci, Massimo. 2009. The End of Theory in Science? *Embo Reports* 10 (6): 534. https://doi.org/10.1038/embor.2009.111.
Roffe, Jon. 2015. *Abstract Market Theory*. London: Palgrave Macmillan.

Rouvroy, Antoinette. 2011. Technology, Virtuality and Utopia: Governmentality in an Age of Autonomic Computing. In *Law, Human Agency and Autonomic Computing. The Philosophy of Law Meets the Philosophy of Technology*, ed. Mireille Hildebrandt and Antoinette Rouvroy, 191–140. London, New York: Routledge.

Rouvroy, Antoinette, and Thomas Berns. 2013. Algorithmic Governmentality and Prospects of Emancipation. Disparateness as a Precondition for Individuation Through Relationships? Translated by Elizabeth Libbrecht. *Réseaux* 177 (1): 163–196. https://doi.org/10.3917/res.177.0163.

Rouvroy, Antoinette, and Bernard Stiegler. 2016. The Digital Regime of Truth: From the Algorithmic Governmentality to a New Rule of Law. Translated by Anaïs Nony and Benoît Dillet. *La Deleuziana* 3: 6–29.

Simondon, Gilbert. 1992. The Genesis of the Individual. In *Incorporations*, ed. Jonathan Crary and Sanford Kwinter, 297–319. New York: Zone Books.

———. 2009. The Position of the Problem of Ontogenesis. *Parrhesia* 7: 4–16.

———. 2017. *On the Mode of Existence of Technical Objects*. Translated by Cécile Malaspina and John Rogove. Minneapolis: Univocal.

———. 2020. *Individuation in Light of Notions of Form and Information*. Translated by Taylor Adkins. London and Minneapolis: University of Minnesota Press.

Smith, Daniel W. 2005. Critical, clinical. In *Gilles Deleuze: Key Concepts*, ed. Charles J. Stivale. Kingston and Montreal: McGill-Queen's University Press.

Srnicek, Nick. 2016. *Platform Capitalism*. Cambridge: Polity Press.

Stiegler, Bernard. 1998. *Technics and Time, I: The Fault of Epimetheus*. Translated by Richard Beardsworth and George Collins. Stanford, CA: Stanford University Press.

———. 2009. *For a New Critique of Political Economy*. Translated by Daniel Ross. Cambridge: Polity Press.

———. 2013. *What Makes Life Worth Living? On Pharmacology*. Translated by Daniel Ross. Cambridge: Polity Press.

———. 2016. *Automatic Society Volume 1: The Future of Work*. Translated by Daniel Ross. Cambridge: Polity Press.

———. 2018. *The Neganthropocene*. Translated by Daniel Ross. London: Open Humanities Press.

———. 2019. *The Age of Disruption. Technology and Madness in Computational Capitalism*. Translated by Daniel Ross. Cambridge: Polity Press.

Stiegler, Bernard, and Irit Rogoff. 2010. Transindividuation. *e-Flux* 14: 1–6. https://www.e-flux.com/journal/14/61314/transindividuation/.

Tynan, Aidan. 2010. Deleuze and the Symptom: On the Practice and Paradox of Health. *Deleuze Studies* 4 (2): 153–160. https://doi.org/10.3366/E1750224110000929.

Varela, Francisco. 1999. The Specious Present: The Neurophenomenology of Time Consciousness. In *Naturalizing Phenomenology: Current Issues in Phenomenology and Cognitive Science*, ed. Jean Petitot, Franscisco J. Varela, Barnard Pacoud, and Jean-Michel Roy, 266–314. Stanford University Press.

Vignola, Paolo. 2017. Symptomatology of Collective Knowledge and the Social to Come. *Parallax* 23 (2): 184–201. https://doi.org/10.1080/13534645.2017.1299295.

Wark, McKenzie. 2019. *Capital is Dead.* London and New York: Verso.

Wiener, Norbert. 1965. *Cybernetics, or the Control and Communication in the Animal and the Machine.* Cambridge, MA: The MIT Press.

Zuboff, Shoshana. 2019. *The Age of Surveillance Capitalism. A Fight for a Human Future at the New Frontier of Power.* New York: Public Affairs.

6

The Surveillance Axiomatic

S. L. Revoy

One of the most notable fissures in the Deleuzo-Guattarian writing machine concerns their treatment of computers and the computerization of society. This divergence forms an ambivalent line of flight within their collective corpus which does not subject the incipient rise of information communication technologies and its transcendent pretences of universality to the same level of critique as the other "lofty objects" (Guattari 2009a, p. 280), the arborescent signifiers of royal sciences claiming to uncover mechanisms governing the structures of organization, which are the focus of their joint works, especially throughout Capitalism and Schizophrenia. This may seem like an absurd claim given the enduring reputation of Deleuze's writing on the control society, which specifically aligns computerization with an emergent post-disciplinary society defined by an increase in the extensity and intensity by which ordering forces are able to modulate society because of the saturation of information communication technologies (Deleuze 1995a). While there is clearly, in Deleuze's control society thesis and elsewhere, a recognition of the

S. L. Revoy (✉)
Queen's University, Kingston, ON, Canada

S. S. Das, A. R. Pratihar (eds.), *Technology, Urban Space and the Networked Community*, https://doi.org/10.1007/978-3-030-88809-1_6

capacity for computers and processes of computerization to engender new forms of oppression, cybernetics, the intellectual font of computers, and their arborescent foundation is not addressed in the sustained and unreservedly critical way in which Freudo-Lacanian psychoanalysis, Saussurean semiotics, or other exemplary exponents of the structuralist movement are treated. Indeed, there is scarcely any explicit mention of foundational figures in cybernetics and the central arguments underpinning its universal science of control are essentially unremarked. This is a curious state of affairs for several reasons. First, I say "essentially unremarked" because, in truth, a litany of oblique, reproachful remarks about information theory and cybernetics punctuate Deleuze and Guattari's (1987, pp. 5, 15–16, 79, 179, 1994, p. 209) joint writing, while Guattari (2009b, pp. 229–230, 2011b, p. 13, 103–104, 2016, pp. 190–192) by himself frequently offers more substantial critical remarks in his solo writing. The absence of a sustained critique of cybernetics by Deleuze and Guattari is therefore even more curious, especially given the contemporaneous pre-eminence of cybernetics and its highly controversial claims as to the discovery of universal mechanisms of behaviour and control across the organic and inorganic. In its famous assertion that the world is comprehensible as a series of interwoven cyclical structures where feedback is constantly assessed by purposive, goal-seeking entities which respond rationally to the sense data of this feedback, adjusting to it accordingly in teleological pursuit of their goal, cybernetics—beyond the ambitions of any other structuralist endeavour—posits that the organization of the world comprises purely arborescent structures animating and cohering their seemingly stochastic, chaotic nature through rational, communicational processes which can be mathematically simulated and theoretically controlled with sufficient technical sophistication and rigorous experimentation. The extent to which this majoritarian strain of cybernetics is anathematic to Deleuze and Guattari's philosophy is difficult to overstate, a point well-reflected in the passing comments which comprise their treatment of the subject as an exemplar of arborescent thought. So far, so consistent. However, in spite of their piecemeal criticisms and occasional mockery of many foundational scientific concepts which form the intellectual bedrock of cybernetics and the enumeration of its conglomerated scientific foundations, like information theory, alongside the objects of critique they single out most trenchantly, there is no elaborative critique of cybernetics as an onto-epistemological model of reality which is corollary to their multifaceted critiques of projects like Freudian psychoanalysis or Chomskyan linguistics. There is a trenchant acknowledgment of the arborescent condition of cybernetics as a virtual model, but no critical inquiry into the problems generated by the concrete application of its principles to the organization of everyday life through computerization.

The divergence in Deleuze and Guattari's appraisal of computers and computerization arises due to two elements of Guattari's disposition towards cybernetics and the use of computers: his belief in the separability of cybernetics from the possibility of creative experimentation to which the technological fruits of cybernetics might be put to work and, relatedly, his highly favourable appraisal of the minor science of cybernetics as put forth by Gregory Bateson. The schismatic intellectual history of cybernetics offers an analytic which is useful in order to initialize an evaluation of the comparative reluctance of Guattari to offer any sustained reflection on the world-historical transformation—of the integrated world economy, of the state-form, of the production of subjectivity—brought by computerization, at least relative to Deleuze. Guattari's recognition that the precepts of the royal science of cybernetics constitute a particularly pernicious form of arborescent logic is clear. He expresses severe reservations about the scientific metaphors employed by cybernetics, such as the articulation of homeostasis as the condition to which all entropic systems aim to return—a foundational referent for the cybernetic feedback loop (see Ashby 2015)—stating that "I do not trust metaphors from thermodynamics. There need not in theory be any need for a closed action/reaction circuit, a return to the original state" (Guattari 2009b, p. 229), but equally obvious in his work is Gregory Bateson's minor cybernetics. One of the formative influences of Guattari's ecosophy (see Guattari 2000), Bateson saw cybernetics not as the closure of epistemology, an argument quite forcefully and literally made by certain radical proponents of cybernetics (see von Foerster 2003), nor as a universal method to determine the structure and control mechanisms of all stochastic activity, but rather as a philosophical mode to reflect upon the way in which elements of an environment (dis)integrate, propagate, or break down, in the combinatory effects of its discrete ecological elements. Bateson's work is a rearticulation of cybernetics without its "rather impoverished mathematics" (Guattari 2016, p. 191) and which instead offers a means to schizoanalyse the relations between components of ecologies in terms of their operative roles. Divested of its pretence of universality and obtuse scientific rationalism, it is obvious why such a theory would appeal to Guattari, and why the royal science of Wiener and Shannon's cybernetics earned his intellectual contempt. This line of inquiry into Guattari's relationship to cybernetics also helps explicate his reluctance to consign computer equipment to such a model as the Deleuzian control society—he remained unwilling to disavow the creative possibility in such technologies and their capacity to offer new means of expression which generate lines of flight and new minoritarian spaces:

> Thus 'computer-aided design' leads to the production of images opening onto unprecedented plastic universes. … But then again, we should be on guard against progressivist illusions or visions which are systematically pessimistic. The machinic production of subjectivity can work for better or worse. … It's impossible to judge such a machinic evolution either positively or negatively: everything depends on its articulation within collective assemblages of enunciation. At best there is the creation … of new Universes of reference; at worst there is the deadening influence of the mass media to which millions of individuals are currently condemned. (Guattari 2006, p. 5)

In assessing the sum of Guattari's comments on computers, the process of computerization throughout societies, and the intellectual foundations of their production, one finds ambivalence verging on stasis, a clear recognition that there is an arborescent horizon which haunts computerization in its onto-epistemological basis relying on "a reduction, a translateabilisation in terms of quantities of information, that is to say, in the last analysis, a structured succession of binary choices that can be treated exhaustively by a computer" (Guattari 2016, p. 190) combined with a refusal to conflate its potential use with any intrinsic value, maintaining that such collective equipment cannot be evaluated absent a pragmatics of the assemblages which utilize the equipment. We may contrast this with the Deleuzian position which is far more closely aligned with

Foucault in recognizing the diagrammatic function evinced by new forms of knowledge, the correlative field of power they constitute, and the actualization of these fields in the compounding of new discursive unities and apparatuses of power. These apparatuses take bodies up on a simultaneously aggregative and individuating basis, disciplines being an exemplary form of this process, to form kinds of subjects, models of subjectivity understood as predicates of their sociocultural milieu and its diagrammatic rendering of ideal participants (see Deleuze 1988, pp. 31–44). While there is always resistance in the societal models offered by Deleuze and Foucault, the line of flight is inessential to its structure and function, whereas for Guattari such societies are always defined by their ruptures and the contingency of rupture.

It is a lamentable truth that Guattari's highly qualified optimism has been unfulfilled thus far, though the dissection of this state of affairs proves edifying as to the special category of problematic generated by the legacy of cybernetics and cybernetic technologies. This retrospective dissection should not be construed as a critique of Guattari because the extent to which cybernetics and its derivative products operate based on an axiomatic system premised upon the arborification and capitalization of information has only been thrown into relief as the computerization of the world has increasingly resulted in its saturation by a particularly egregious, seductive, and invasive form of topological surveillance which has rapidly become a constituent, normalized element of everyday life. Understanding why the foreclosure of Guattari's hopeful reservations about the creative possibilities for computer technology is thus far characteristic of the state of affairs where computerization is concerned requires consideration of the lineage of cybernetic media vis-à-vis the following: a cartographic, non-anthropocentric approach to surveillance assemblages which is divorced from contemporary received wisdom as to the function of surveillance which is derived from the preponderance today of sociological and juridical explanations regarding computerized surveillance; the application of such a cartographic approach to the rearticulation of the royal science of cybernetics and its media apparatuses on as representative of an axiomatic science of surveillance; and to consider the integration of this surveillance axiomatic into the world capitalist system. The persistent force exerted by technologies designed according to the principles of majoritarian cybernetics—and there are precious few examples of computer technologies which have not employed these principles wholesale since the postwar emergence of cybernetic design as a going concern in academia and industry—is functionally axiomatic. The unique manifestation of cybernetics as a concrete force of order lies in the instrumental medium of the computer, itself a tabula rasa which operates as a "functional analog to ideology" (Chun 2011, p. 59), allows its concrete machines to reify the ideal of its axiomatic structure. As an abstract machine, cybernetics is doubtless the most influential innovation in the history of penal technologies since Bentham's panopticon in terms of its genealogical legacy—it is the diagrammatic function of the control society's Urstaat, the universe of reference which founds the state-form of dividuating control. Utilizing Deleuze and Guattari's critique of the capitalist axiomatic is invaluable in both elucidating the axiomatic structure of cybernetics, its technologies and order, as well as its isomorphic interface with integrated world capitalism.

Cartographies of Surveillance

Four elements organize, cohere, and animate the content and expression of a surveillance assemblage: sense, desire, knowledge, and vigilance. Retrospectively evaluating an arrangement of forces within these terms reveals two important initial points: taking each term as decidedly non-anthropocentric, the cycle of assemblage as such cannot be dissociated from a certain level of surveillant activity as a predicate of its assembling; second, that the virtual concentration, relative concretion, and types of combinations between these terms both virtually and actually suggest cartographies of surveillance which are infinite and vary wildly even within primarily anthropogenic cases of surveillance. In each case we will provide a definition of the term within the context of surveillance cartographies as well as examples which demonstrate its variable function within analogue situations alongside its arborified function as part of the surveillance axiomatic governed by cybernetics and cybernetic media.

Sense

Sense is both the expressible or the expressed of the proposition and the attribute of the state of affairs. It turns one side towards things and one side towards propositions. But it does not merge with the proposition which expresses it any more than with the state of affairs or the quality which the proposition denotes. It is exactly the boundary between propositions and things. It is a liquid at once extra-Being and inherence, that is, this minimum of Being which befits inherences. It is in this sense that it is an event: on the condition that the event is not confused with its spatio-temporal realization in a state of affairs. We will not ask therefore what is the sense of an event: the event is sense itself (Deleuze 1990, p. 22).

The production of sensible propositions which may be correlated within a knowledge system is the task of surveillance assemblages. The successful combination of the wasp and the orchid—the exchange of pheromones, the coherence of their symbiotic arrangement—is a productive outcome of a surveillant assemblage as much as anything else, a cohering of the sensible as a predicate of the knowledge leading to pollination. Surveillance actualizes, through highly variable mediatory practices, the partial propositional capture of sense, organizing and enunciating its sense effects according to the criteria by which it coheres and differentiates sense from nonsense and articulates its findings in relatively virtual—such as inscribed in memory—or relatively concrete ways, such as calling out to an intruder. The undifferentiated plane of sense and nonsense is the matter of surveillance assemblages, that which is somehow sensed, depending upon the surveillance media being employed, and which is relayed in propositional form which is discursively legible—another atomic element of particular systems of knowledge with correlate relations of political force. Someone drawing a firearm after finding an intruder in their home, which happens to be located in a jurisdiction which has legalized the use of deadly force in cases of home invasion, has correlated their sense of the situation, their fixing of the events unfolding, with the apparatus of power interlacing the gun rights discourse and juridical force.

Sense, being a purely ideational event in the Deleuzian model, can only be partially apprehended and articulated in propositional form, forever deferring access to pure sense in the inherent limit of representing the sensible through recourse to some sort of linguistic order—including the performative order of non-anthropomorphic examples like the orchid and the wasp. The articulation of propositions and the precursory capture of a particular sense of sense, the attempt to fix an event as a known quantity and thereby enable expressive articulation, are manifest qualities of both the axis of content as well as expression in any surveillance assemblage. The contents of a surveillance assemblage suggest the types of sense which the assemblage is oriented towards fixing and the knowledge it seeks to gain, while the elements of expression suggest the collective equipment which serve to capture, parse as sensible, and rearticulate as proposition the sensible objects towards which it is oriented to surveil. A family dinner table is a hardened sort of surveillance assemblage, its contents the relations between family members and the political structure of the family, the emotional states of all involved and all the little machines they bring with them to that particular meal, while the equipment of expression involve the order of language, sightlines at the table, the possible multimedia experience of some being intoxicated by alcohol, as much as sense organs and the surveillant cognition of the brain. Perhaps the father, playing the patriarch, wants to "get a sense" of how his children are doing in school. He peppers them with questions and rearticulates their answer as propositional forms in accordance with his own criteria, evaluating their responses to such rearticulations, and repeats this interrogatory cycle, each time attempting to draw out and capture a particular sense, each time deferring any true capacity to capture this sense through the limits of language and his own subjective take on the state of affairs of his children. Thus this insular, local surveillance assemblage both takes and produces sense as its object. This does not need to have a disciplinary or otherwise political valence: a smoke detector's signal, itself acting as a cybernetic surveillance assemblage unto itself with its sensory mechanisms piloted by a simple computer circuit programmed with a binary threshold, disturbs one's sense of the domestic sphere and the surveillance assemblage which integrates home and subject—perhaps a particularly hyperactive assemblage, if one is of a despotic or paranoiac disposition and constantly awaits threats from the outside, noisy neighbours or trespassers to be confronted.

Knowledge

We can say as much for the subjects and concepts of a statement. A proposition is supposed to have a referent. That is to say that while reference or intentionality is intrinsic and constant in propositions, whatever fulfils that purpose is extrinsic and variable. But this is not the case with statements: a statement has a "discursive object" which does not derive in any sense from a particular state of things, but stems from the statement itself. It is a derived object, defined precisely by the limits to the lines of variation of the statement existing as a primitive function (Deleuze 1988, pp. 7–8).

The domain of knowledge, of discursive formation and citation, is the crux of differentiation between the passive surveillance assemblage of one who is at peace in their home and unbothered by the activity of the outside and the paranoiac despotic subject treating that same house, with the same situation relative to the outside, as a jurisdiction to be strictly policed and prosecuted. Knowledge, like sense, is both a referential domain and future object of a surveillance assemblage which serves to filter and operationalize sense as an element placed in relation to discursive formation. Here we use knowledge in the classic sense of Foucault (1980, pp. 111–114): as a regimentary force of order whose establishment as truth, as legitimated knowledge, is the effect of a discursive unification which finds successful uptake, defined by its acceptability as legible and legitimated truth, in the world, both theoretically and concretely, through reference in the non-discursive space of practices as much as in its continual reinscription as truth via multivalent citation, its ongoing discursive formation. The production of knowledge necessarily invokes the creation or maintenance of a regime of truth and a correlative field of political effects which reference discourse while sustaining it in the efficacy of its political articulation; the vile cyclical contrivance wherein minorities, especially black men, are incarcerated at wildly disproportionate rates due to preconceived racist biases regarding the criminality of minorities, and whose disproportionate incarceration is then rendered as statistical evidence to support the empirical truth of this pattern and set of preconceptions, is a readily accessible and catastrophic example of this reciprocal relation between the discursive field of knowledge and its correlative field of concrete political effects. The bureaucrat or legislator looking to justify systemic racism in the penal system forms its own epistemic surveillance assemblage, along with their staff and donors and all the machines working upon them, casting about for information or data which may be rendered unto their cause as sensible to it, animated in part by their own constitution within the pre-existing discursive formations of systemic racism and who aim to collate and organize new elements to continue its effective power, eventually formatting and articulating some of the effects of this regime of truth in statistical form as prima facie evidence of its legitimacy as a form of knowledge about a particular deviant population which governs political action. The field of knowledge serves a precursory function in the animation and orientation of a surveillance assemblage. The discursive formations which it draws on orient its criteria of analysis—consider the many different things the father at the table might be concerned about depending upon his unique neuroses and disposition towards his children and the lives of children more generally, as well as his sense of phallocratic sovereignty. The productive outcomes of any surveillance assemblage can be formulated as the production of sense as it is parsed and articulated through the domain of knowledge, generating new discursive elements to potentially be articulated within a given regime of truth. Knowledge thus filters, organizes, and manifests the production of sensible propositions as they relate to pre-existent regimes of truth which are cited in the organization and articulation of any surveillance assemblage. This is the case for both the components of content and expression in a given assemblage. A teenager being allowed to go to a dance by their watchful parents is diagrammed within a surveillance assemblage which is predicated and premised upon the discursive conditions in which it is organized and produced: at certain places and points in history, such is the regime of truth corresponding to teenage sexuality and social mores that a chaperone would be de rigueur, whereas today it may be a constant line of contingent control via cellular communication. Both the contents, the formed matters making up the material situation and the expressive components representing functional relationships, in this surveillance assemblage are mutable and historically situated, but the function of knowledge and its correlative field of power is not, nor is its productive effects. Surveillance as the diagrammatic function of assemblage, of processes of assembling, relates to desiring-production and the genesis of any particular organism as a metastable construct as an emergent, concrete form which separates out of the undifferentiated full body of desire, the "production of recording," while surveillance as a particular type of assemblage premised upon discursive productivity relates to social production (see Deleuze and Guattari 1983, pp. 7–16). In both cases, which are obviously not divisible in that the latter always necessitates the former (any surveillance assemblage is at minimum a twice-articulated form of surveillance), the phenomenon of surveillance represents the action of recording which Deleuze and Guattari use to conceptualize the extraction of metastable unities from the undifferentiated fields of antiproductive matter constitutive of desiring-production (desire as such) and social production (the full body of the socius). A father, animated by micro-fascistic desires to control his home and family in the homogenous terms which are harmonious to his neurotic self-image, sees his daughter wearing clothing of which he disapproves; one can almost hear the click of the equipment as the recording process begins. Buttressed by his privilege, his phallocratic certitude about proper parenting as much as proper young women, he is empowered by his own regime of truth—one well-reflected in his life and experience of reality, no doubt—to cast judgement, to form a rapid and crystallized sense of his daughter's appearance, information immediately placed into relation with the discursive unities in which he is constellated. In his repressive actions, his prohibitions not only constitute micro-fascistic desire, nor a disciplinary encounter, but the proliferation of certain statements and neutralization of others which cohere the discursive formations which he implicitly cites in his prohibitions—his repressive gaze, caustic comments, interdictory commands all form part of the microphysics of discourse which help stabilize its power through citation of its truth-value and the successful enactment of assemblages making use of them. Thus, a certain social relation, a repressed figure of the daughter subject, a phallocratic father figure, as well as the cohesion of discourses around protective dads and improper dress and teenage repression, are all recorded and reinscribed in the encounter. This cannot be dissociated from the surveillant dimension of assemblage as such, and in this case (as in many) from explicit consideration of conditions of surveillance assemblage.

Vigilance

It is not the slumber of reason that produces monsters, but vigilant and insomniac rationality (Deleuze and Guattari 1983, p. 112)

Owing to their special status as components of a surveillance axiomatic, computers have not yet been featured in a starring role in our discussion of surveillance assemblages—which still, in the examples given, involve the conjugation and arrangement of codes, relative decoding and occasional irruptions of overcoding, rather than the absolutely deterritorializing action of decoding which characterizes an axiomatic system, indeed the computer as such. But here it must be said: computers are vigilance machines; it is perhaps their most unqualifiable power of surveillance. Vigilance is a term which, used here, is proper to behavioural ecologists but which finds itself circulating in other forms across disciplines; in psychology it is given the more individuated term task saliency. How long can something be done? How long can activity be maintained? These are the questions of vigilance, and the concrete force of coherence in a surveillance assemblage. A stalker is a highly disturbing figure for multiple reasons, but their extraordinary force of vigilance is an important element. Being followed down the street for any length of time is by itself terrifying because of the contingent threat of further transgression, but also because of the deferral of any recognizable terminus: how long will this go on? The more a stalker reappears, in different places, despite police interventions, the more it suggests not only a concerted desire, an intensive level of knowledge regarding the victim (and a correlatively generated field of political power, as when new knowledge of a home address leads to a new instance of stalking), but an unnerving force of vigilance, a concrete cohesion, articulated materially through dopamine as much as libidinal investment and micro-fascistic enticements to dominate, of a desire to keep the surveillance assemblage alive. However, the most obsessive stalker, the most vigilant predator in the wild, is nothing in comparison to the vigilance of a computer. Absolutely programmatic, requiring only electricity and the maintenance of its mechanical parts, the vigilance of a computer is theoretically limitless in comparison to the finitude of humanity, which is why its capacity is not measured in how long it is able to do something, but how long it will take to do anything given the amount of processing it can constantly accomplish. The measurement of clock-time as the inferential metric of a computer's processing power contains the presumption of an infinite vigilance under normal conditions of operation. In each case, vigilance has an important functionalizing role as a quality of the content of surveillance assemblages, quite literally determining in this case how intensive and durable the action of the assemblage can be, and in the expressive components—excessive vigilance can have an affective dimension which is correspondingly creepy or even terrifying, as in a stalker or even in someone who stares excessively.

Desire

Desiring-machines make us an organism; but at the very heart of this production, within the very production of this production, the body suffers from being organized in this way, from not having some other sort of organization, or none at all (Deleuze and Guattari 1983, p. 9)

Desire, rather desiring-production, designates at once the most local coordinates of a surveillance assemblage and the broadest organizational field within which such assemblages are merely pieces, conjugations which convoke other abstract machines—we have already mentioned the relation to discursive unities which surveillance assemblages necessarily entertain, at once in thrall as well as a potential constitutive force, and one such example. There are likewise two indeterminately reciprocal articulations of the surveillance assemblage within desiring-production: as a product of this process and as a crucial organizational vector of desiring-production as such. The former is likely far more easily comprehensible in the sense that one can imagine innumerable scenarios where the desiring-production inherent in all concrete machines makes productive use of surveillance—consider Foucault's disciplinary society and the variety of techniques and goals which form the surveillance assemblages within different disciplinary enclosures. Foucault's diagnosis of the disciplinary society as an apparatus of capitalist subjection should also alert us to the second manifestation of surveillance assemblages which do not aim to produce any particular object due to the operationalization of surveillance as such, but which functionalizes the organization of particular abstract machines due to their place within the order of more intensely saturated abstract machines (see Foucault 1980, p. 221). This represents a doubling of the twofold articulation of surveillance already outlined, where surveillance is both a force of organization for assemblage as well as a type of assemblage which organizes and produces on the basis of surveillance. Further to this, it is also the case that certain abstract machines which achieve a systemic reach and saturation in the world may also coordinate surveillance assemblages on the basis of a diagram which simultaneously employs surveillance as an organizational vector while also diagramming the productive ends to which the actual surveillance assemblages may be put to work; integrated world capitalism is the exemplar par excellence here. While each disciplinary enclosure in Foucault's model forms surveillance assemblages on the basis of their particular goals—producing model students, pacified and shameful prisoners, children conforming to a phallocratic social structure—each of these facets of the disciplinary society is coordinated within a broader, unarticulated schema premised upon producing dependable and well-trained workers for a capitalist economic system where all the lessons of different enclosures form a set of subjective qualities which allow a worker to be flexibly integrated wherever they happen to land in the economic order. A more local, quotidian example: searching for my wallet, keys, hat, boots across my apartment does not engage a surveillance assemblage in service of a desiring-production of surveillance, but as an organizational vector for gathering my daily outfit, itself absolutely the product of a process of a sartorial diagram representative of a subjective process within desiring-production vis-à-vis my conscious design of myself in relation to the milieu of contemporary fashion. It is often the case, as in the disciplinary society, that these articulations exist in a formal reciprocal movement, as in capitalism and discipline, but this is not a universal condition of surveillance assemblage as either a vector of desiring-production, a mode of its organization as assemblage, or in the surveillant miraculation of desiring-production in which organization of any assemblage occurs on the basis of surveillance as such—not as mere instrumental function of a particular machine or constitutive of a particular desired effect, but as the machinery animating the entire imaginary of desire and its machination. Again, these are not necessarily separable phenomena: some surveillance assemblages participate in other abstract machines alongside their own explicit aims involving the use of surveillance—regardless of his idiosyncrasies, all tyrannical fathers pay homage to the regime of phallocratic domination in their familial surveillance assemblages—while some are actualized as the cohesive force organizing an assemblage, articulating an abstract machine using the instrumental utility of surveillant actions, like hunting for items in an apartment, without surveillance being the organizational locus or font of desiring-production. The former is a surveillance assemblage, the latter is the surveillant dimension of the diagrammatic function which organizes any assemblage. All organizations, regardless of style (and all order with a particularly paranoid obsession), rely upon the surveillant dimension of the diagrammatic function to maintain coherence. It is the mechanism which draws line, makes connection, it is the function which does the actual diagramming. We know ourselves only in constant reference to our understanding of that which makes us who we believe we are, and the same goes for any group subject, social segment, or society. It is a constant recursive exercise where the diagrammatic function of a particular identity as it is superimposed on a concrete entity, whether person or nation-state, requires the renewal and reinvestment of the diagram through a survey and affirmatory ordering of the elements of the identity. The relentless interdiction of the superego is representative of the repressive facet of this dimension of the diagrammatic function as it is found in Freudian subjectivity, scanning the contents of the mind and sense of self for evidence of psychic malfeasance. A less grim example: aesthetic experience, like sitting on the shore of a lake, requires the instrumental function of surveillance in order to occur, but it is not a case of placing the lake under surveillance with some sort of productive aim derived from the instrumental function of surveillance—this is the case with a phenomenon like resource extraction, which does place parts of the earth under surveillance. The productive effects of aesthetic experience are incidental to the powers of the surveillance assemblage, which still coordinates the four outlined elements in the same cartographic manner, when it is manifest as the organizational force of a diagrammatic function, actualizing the virtual, instrumental elements (cognition, sensory function, mental reflection, attention, etc.) necessary to experience the shore of the lake, as it is when it is a concrete surveillance assemblage whose desiring-production and concrete existence are structured by some form of organized monitoring which has productive aims correlating to its particular assemblage and the surveillance it is premised to undertake. It cannot be stressed enough that these two articulations exist in a contingent condition of indeterminate and reciprocal transformation; a man sitting on a park bench surveying the weekend picnickers and runners is engaging in surveillance insofar as the diagrammatic function of his roving, unfixed gaze and its itinerant connection to all elements of the park is concerned, but as soon as he begins to stare and track someone with a particular interest in following them, gazing longer or worse, he is no longer simply part of the park assemblage, but has inflected his gaze with other elements, especially a newly enunciated desire, motivating its cross of a threshold into a surveillance assemblage which uses the force of surveillance in a self-possessed, purposive manner to achieve a productive end which is proper to the function of a particular sort of surveillance. The former is schizoveillance, an unanchored gazing without discursive rationality and a desiring-production of the gaze which is unfocused on any particular elements of the park, its visual sense of the world a kaleidoscopic, rhizomatic survey; the latter is surveillance assemblage, establishing arborescent concretion as dictated by its mutated coordinates of purposive, fixed monitored, molarization of the desiring-production of the gaze and the correlative constitution of a field of knowledge—in the case of a man staring at an attractive woman without regard to her comfort, it is the reconstitution of the discursive condition of male privilege, convoking a regime of truth regarding masculine liberty with the desiring-production of a self-consciously privileged subject as they organize and express sexual desire unilaterally through their gaze and all the abstract machines sustaining phallocratic dominance—molar social strata functionalized through surveillance assemblage with the productive aim of conspicuous objectification and the reinscription of male privilege, writ micropolitical.

Running in the Forest: Cybernetic Media and the Surveillance Axiomatic

The quest for "universals of communication" ought to make us shudder (Deleuze 1995b, p. 175)

The surveillance axiomatic in which we exist today is the twofold articulation of the logic of majoritarian cybernetic taken to its logical extreme. It is first articulated as a royal science, instantiating its virtual condition long before the technical sophistication existed for computers to make manifest its totalizing ambitions to simulate the stochastic. Cybernetics has always entertained a hard core of fanatics who believe, as crystallized most brazenly in the work of Heinz von Foerster, that the discovery of cybernetics encircles, encloses epistemology by offering the skeleton key to all understanding. Three binarizing axioms converge to promote this vision (Pias 2015). First, that all of the world communicates in terms of fundamentally binary messages—the signal and the noise, and that these signals can be deduced through sufficient surveillance technologies, sensory prosthetics allowing the system in question to be analysed, and with sufficient analytical capacity, that is, programmability (Shannon's information theory) and the electrical engineering innovations positing a functional model of the cybernetic onto-epistemology in the computer medium itself, a medium of purely cybernetic mediation. Second, that all such systems exist in a closed processual loop involving a series of arborescent logical concepts like feedback, which posits that any element of any form of organization exists in a state of monitoring for feedback in terms of the elements in which it is constellated while attempting to attain its goal and modulates its behaviour on the basis of that feedback (Wiener's cybernetics). Finally, that manifested behaviour which is observable in a cybernetic system is necessarily an exercise in rational calculus which moves with constant purposive intent towards a goal in a teleological fashion and which can therefore be rendered in a sequence of binary terms aligning the analysis of behaviour with the signal/noise binary of information theory and with purposive/non-purposive behaviour (Rosenblueth-Wiener-Bigelow model of behaviour), readily translatable into the binary machine language of digital computers. Much of the conceptual language and philosophy of cybernetics has been discarded in the development of computer science and its eventual schismatic compartmentalization of different aspects of computer design, such as interface design, as separate elements, but its precepts remain foundational, the essence of the discursive condition of computerization, a set of axioms dictating its limitless capacity for simulation. Considered as a disembodied, virtual condition of computerization, the statements of the royal science of cybernetics coalesce around the promotion of the concept that any phenomena, however complex, can be simulated, modelled, and theoretically controlled through a sufficiently sophisticated cybernetic construct with the means to monitor the system in question, its many feedback loops and communicational elements, as well as the means to capture and analyse the resultant information as computable data. Taken as an onto-epistemological model, the royal science of cybernetics represents the world as a forest, an absolutely arborified world whose rhizomatic dimension is merely an inadequately analysed stochastic system which will always be revealed to accord with the fixed structures of the cybernetic model of reality if it is properly computerized, rendered in accurate simulational terms where its underlying systems of cybernetic governance might be calculated, revealed, and made susceptible to intervention and control. The computer is, as Wendy Chun has correctly assessed, the functional and functionalizing mechanism of this model—it both provides an ideal type of the cybernetic system in its design and function while also serving as a medium by which analogue phenomena may be modelled by the computer, converting the phenomena into a simulational strata where it is made to accord with the cybernetic onto-epistemology, rendered in the complex of binary sequences necessary for a computer to process it as data.

Let us bring this virtual condition down to Earth, two Earths nearly a century apart. Claude Shannon, a figure emblemizing the idea that "every good scientist is half B.F. Skinner and half P.T. Barnum" (Stern 1993), demonstrates his maze-solving machine at the formative Macy

Conferences on cybernetics. The maze raises and lowers walls at random while an electronic mouse attempts to navigate it. After each attempt, the computer running the mouse notes how it was stymied and attempts a new path, quickly computing a probable and successful path through the maze in spite of its dynamic reconfigurations. Such is the world according to the onto-epistemology of cybernetics: a series of rational, goal-seeking endeavours which can be simulated and optimized with the proper means of capture, analysis, and control over the environment and flows of signals, each process governed by the processing of different sorts of communicative feedback inputs which can be analysed to assess the optimal path towards a given goal (Shannon 2015). In 2018, Mark Zuckerberg is summoned the Congress of the United States to answer some questions about his particular maze machine. Zuckerberg's maze is the human subject, his machine a network which ostensibly simulates the social as such. Rather than running into walls, Zuckerberg's machine captures all the communicational elements of its users that it can—their self-expression, private discussions, other browsing habits while using Facebook, relative level of interest in different users, major life changes, shopping habits—relentlessly inferring the subjective state of its users in order to exert capitalistic control over them as such through targeted advertising and the modulation of behaviour to an optimal state, and to liquefy their captured data as market research to be parsed, packaged, and wholesaled. Zuckerberg's maze machine does not require a single computer, but billions of computerized devices, each serving as a node through which intense, constant surveillance is conducted. These are the sensory prostheses which enable Facebook's simulation of the subject and the social, inducing through a complex of alternately disciplinary, controlling, and seductive forces the participation of users in this massive surveillance network premised upon the simulation and commodification of the social and the subject (Terranova 2015). That Facebook is far more adept as an engine of commodification than simulation is not surprising, as will become clear as we elaborate the parallelism of the surveillance and capitalist axiomatics, but its success in this regard is irrelevant. The dominance of Facebook, and the course of the networked computerization of the world, demonstrates the untrammelled success of the royal science of cybernetics. There is no phenomenon so complex, unweildy, unstable, or dynamic that it cannot be conceived as being subject to computerization and simulated representation through cybernetic media in the current discursive regime of computerization. Today, it is largely the users of such technologies who constitute the black box system under investigation even as they themselves use cybernetic technologies to accomplish daily tasks in their lives, a perverse inversion of the classic image of the scientist piloting a cybernetic system in order to unravel the mysteries of some stochastic system. Rather than the increase in knowledge and validation of cybernetics, the purposes of these oft-literal experiments are to increase the appeal of such simulational worlds so that there is more commodifiable data generated. Regardless of change in the political impetus of such machinations, their sophistication, or the complexity of the phenomenon, it is the cybernetic onto-epistemology and its archly arborescent closure of the world which form the diagrammatic function guiding the metastability of both the virtual condition and concrete machinery through which Facebook is actualized.

It is not just that computer technology is well-suited for surveillance, it is that the entire scientific enterprise grounding its conditions of use and technical invention is premised upon an absolutely arborescent philosophy of organization which renders everything in terms of a universal informational substance which can only be accessed through a particular form of electronic surveillance which the digital computer is designed on an onto-epistemological level to accomplish and functionalize. The digital computer does not simulate the world, it simulates the cybernetic conception of the world in a way which is contrived to represent whatever it simulates as yet another ideal type of cybernetic reality; anything computerized is thereby represented in the ideal terms of cybernetics, which is to say in terms of a static economy of surveillance and control. The scandal of our age is less that computers continue to be put to the applications they were premised upon, but that the investigatory instruments of cybernetics are now apparatuses of cybernetic control as well, with any use of a computer generating a contingent relation of asignified, multilateral surveillance depending on the particular context of the computer use. In tandem, the success of the royal science of cybernetics and its functionalization through the proliferation of digital computers as the means of organization for all domains of governance, and increasingly of quotidian participation in civic life, have institutionalized a surveillance axiomatic as an inescapable force within the order of the world in which we are all subject to an isomorphic relation.

Axiomatic systems are a special category of arborescent constructs in Deleuze and Guattari (1987, pp. 452–473), defined by their unique modus operandi of decoding all flows according to their axioms because of the transcendent and universal axioms by which they filter and rearticulate that which they treat. Because the cybernetic onto-epistemology presents a virtual condition premised on the idea that all organization is always already a binarized, linearly progressive world of purposive and non-purposive behaviours, signals, and noise in each communication, and cycles of feedback and adaptation operating in service of the homeostasis of the system and the staving off of entropic decay, all domains of cybernetic concern are already approached as if they contained the prefabricated elements of any cybernetic system and merely need to be decoded for these elements, observed, analysed, and finally controlled. The scientific validity, efficacy, or accuracy of such decoding operations is as irrelevant to the saturation of the surveillance axiomatic as the rational allocation of monetary capital is to the function of the capitalist axiomatic—it is a mythological convenience meant to obscure the ongoing methodological elaboration of the "sleight of hand" (Guattari 2011a, p. 13) which is used to project the totalizing image of axiomatic systems, like cybernetic mediation or capitalism, and assure the universality of its decoding operations. Owing to their unwavering (and onto-epistemologically predicated) preconception of the world as an arborified domain of its own axiomatic system, the self-fulfilling prophecy of the royal sciences, we are told by Deleuze and Guattari that functional axiomatic systems only obey laws and confront limits which are immanent to their insular, occlusive logics and serve to maintain their saturation of the world through the addition (and occasional subtraction) of axioms which are necessary to continually expand and maintain this saturation of the world and the representation of transcendent majesty. Overcoming the tendency of the rate of profit to decline, capturing the emergence of new markets, creating new axioms of social welfare to mitigate crises of the market are all core examples in Deleuze and Guattari's model of the capitalist axiomatic. The surveillance axiomatic operates in a strikingly similar fashion, creating new technologies of surveillance in order to extend and intensify the reach of cybernetic mediation, a process which now in some cases has come to remediating infrastructure and the entire apparatus of urban governance in order to capture, simulate, and exert remediatory control over the entire urban spaces under the designation of "smart cities." More locally, the proliferation of smart devices and the Internet of Things index this process of extending and intensifying the relation between quotidian life and the surveillance axiomatic to incorporate the domestic space entirely. Privacy rights and other ethical reform axioms are introduced to mitigate crises of particular products of the surveillance axiomatic in order to stave off any tendency for the decline in user engagement with a particular surveillance technology. Computer hardware advances allow not only for the miniaturization and further saturation of the surveillance axiomatic into the fabric of everyday life, a crucial necessity for the ongoing maintenance of capitalistic dominance more generally, but for the increased analytical capacity inherent in the increase in computational speed, data storage, and network communication infrastructure.

It is not merely on a discursive or technical level that the surveillance and capitalist axiomatics coincide. Their remediatory effects on the pre-existent power formations which form an isomorphic, often profitable, affiliation with them and are thusly transformed by the internalization and valourization of their arborescent logics and fixed, capitalistic systems. Indeed, beyond the monetization of data, the surveillance axiomatic engenders a whole new typology of capitalistic entities and forms of surveillance capital arising from the suffusion of the everyday with mechanisms of control whose instrumental function is a matter of technical knowledge and little else—hence the alluring, frequently mythologized figure of the hacker and the powers of state they can seemingly marshal for their own devices because they know the exploits of the computer system in question. In new cyborg subjectivities, in the rearticulation of old centres of power through the logic of cybernetic mediation, in analogue forms being digitized ad absurdum, the surveillance axiomatic—apprehended always between its virtual condition and reciprocal, concrete machines—represents the attempted superimposition of computerization and the cybernetic onto-epistemology on the entire field of antiproduction. The fact that there is nowhere near the technical sophistication to accomplish such a task, or the fact that the molar rationality of cybernetics and computerization precludes the capture and representation of anything other than its own tautological onto-epistemology, is immaterial to the incursion upon the field of antiproduction which has become the aim of this emergent axiomatic system, bringing capitalizable rationality to the entire undifferentiated bodies of desire and the socius. Even that which exists in opposition to the surveillance axiomatic is, like anti-capitalist economic systems within the capitalist axiomatic, brought into isomorphic relation with the singular cybernetic medium of a computerized world as a condition of civic, social, political, and economic participation in society as such. Anticipating the connection between phenomena like the Internet—the central exchange network of the surveillance axiomatic—and the global market system, Jodi Dean (2009) has compellingly argued that the use of networked computer technologies is indissociable from participation in the commodification of information as data. Instead of fungible forms of money, it is the fungible substance of digitized data which is used as a universal form of exchange, minted and circulated through cybernetic mediation and concurrently treated by the capitalist axiomatic, quite efficiently in fact in that cybernetic media is premised upon the capitalization of information as data. We would also add that, again, alongside monetary capitalization there is an under-theorized typology of surveillance capital which is generated and circulated both as part of processes of financialization but simultaneously in excess of purely monetary valuation.

Becoming-Control

The recapitulation of the surveillance axiomatic in terms of our cartographic approach to schizoanalysing surveillance assemblages reveals the diagrammatic function of the control society. When contrasted with the examples under consideration during our initial delimitation of these four elements, their arborification when applied to the surveillance axiomatic and its process of cybernetic mediation is notable. It is a foreclosure of sense and desire as dynamic or generative, it is a sociotechnical metering of vigilance so that it achieves a state of constancy, a threshold of absolute deterritorialization which attempts to represent the cohering force of the surveillance axiomatic—the mesh of data processing and computerized treatment of worldly phenomena—as a cosmic force of organization, made unthinkable that it would ever cease to exist or surveil, a newly unearthed natural law. It is the disciplinary functionalization of knowledge, articulating the discursive condition as an algorithmic mechanism constantly hailing the user subject to further interact with various, highly seductive, and convenient digital technologies while monitoring with equal constancy the ways by which the user interacts with them, thereby incessantly generating new correlative fields of knowledge and power as its programmatic sense of "knowing" a user triggers potentially far-flung processes based on these evanescent micro-regimes of truth, generated in each databased connection, which may suggest certain advertisements in the most banal outcome, and many other far more nefarious ends besides. The arborification of the discursive function in this case relates to the programmed nature of the interactions generating and conjugating the field of power/knowledge relations, always moving towards a particular outcome through an absolutely grammatical interface medium, and further in the disciplinary relation installed in this discursive relation as the surveillance axiomatic constantly adjusts to be more enticing and better able to capture new forms of data based on user activity. It is the superpanoptic not only in the extensity of monitoring and evaluation, of a modular discipline called control, but in the intensity, the constancy, and multivalence of the experience of control across the topological space of the surveillance axiomatic. Despite the emergence of new qualities defined in the control society thesis, it is the constant evaluation of user input in order to form new knowledge/power relations against fixed criteria which are bounded by intangible fields of enclosure (e.g., a specific website tracking one's purchases) which causes us to argue that the diagrammatic function of the control society retains distinctly disciplinary characteristics even as it turns towards the topographic and the topological and even, to exceed Deleuze, beyond the institutional failure of the disciplines in favour of control.

It is dividuation which most clearly delineates the disciplinary and control societies, a boundary exemplified in the structure and function of the surveillance axiomatic. If the disciplinary society aims to produce individuals who violently internalize the normative ideals of their particular society through enclosure in punitive institutional settings which integrate such internalization of such ideals into their routine function, then the control society aims to draw out and externalize, or dividuate, the individuated subject so it can be assessed and further managed in perpetuity by a multilateral and ever-shifting array of capitalistic powers that share a common, topological network of surveillance and control technologies which are intimately and constantly connected to the subject and, latterly, their environs. It has always been the stated goal of cybernetics to establish a science of control, and in this its heirs have been far more successful as such than in fulfilling the putative intent behind it originally: to understand that which it seeks to control. Nothing on Facebook has ever resembled a person, nor does Facebook Inc. actually know any of its users as whole people, but it is incredibly deft at organizing dividuated profiles of its users through a convenient service platform styling itself as an improved simulational social space, one which is oriented towards the capture of commodifiable information. Facebook is only one node, an especially successful exemplar using a host of common infrastructural technologies, of a topological megastructure which Benjamin Bratton has diagrammed as The Stack, which is constitutive of an axiomatic system of surveillance.

The foundational immanence necessary to construct an axiomatic system of surveillance can be traced to the laws of cybernetics as they were formulated and promoted in its royal strain: totalizing, transcendent laws governing the universe whose limitations were only in our capacity to apprehend complexity. The incredible, epoch-defining invention of the digital computer, a functionally equivalent medium capable of simulating this onto-epistemological construct, enables the concrete realization of its vision of reality in the form of a diffuse topology of cybernetic media which remediate analogue phenomena in a model according with the laws of cybernetics. All experiments in cybernetic control necessitate inventive and intense surveillance ecologies in order for the stochastic system under consideration—recall Shannon's maze—to be unveiled and made controllable. It is the entirely unsurprising, though nevertheless disturbing, end result of the enthusiastic propagation of digital computers, a situation which cannot be separated from its highly efficient augmentation, its twinning of the capitalist axiomatic, which enables a state of affairs in which increasingly inescapable swaths of everyday life are being remediated in the functional terms of a cybernetic experiment, vesting control over it to whomever is capable of gaining access to the parts of the surveillance axiomatic which functionalize that particular topological segment of the control society. Apprehending the surveillance axiomatic is an analytical task as unweildy and multifaceted as the critique of capitalism, and here we have merely begun to sketch its structure and functions as an organizational and orderly strata of the nascent control society. Grasping its unique qualities, the special problematic it represents, necessitates a radical revision of the philosophy of surveillance as it has been theorized, both to unveil the understudied richness of this phenomenon as such and in order to do justice to the molar weight, the arborescent ambitions, of the cybernetic machines which today dominate our world.

1. The royal form is used by Deleuze and Guattari to denote that which acts in a fashion presuming itself to be the conjugal centre of a particularly invariant, hierarchically arranged system, as in the court of a royal. A royal science emblemizes the occlusive, linear, hierarchical, and static image of thought they call arborescence in its rendering of the world as subject to a royal law of invariant, hierarchical, and rationalized structure. See Deleuze and Guattari (1983, pp. 209–215, 1987, pp. 125, 234–235, 361–362).
2. Throughout their discussion of the nature of arborescence as the image of thought and mode of organization privileging stratification, the establishment of hierarchy, and closure of dynamic activity in favour of systems designed around principles of predictable, homogenous structure privileging the stabilization of control, Deleuze and Guattari make several indirect, highly critical comments regarding certain fundamental scientific discoveries underpinning cybernetics. In their

definition of the brain as defined by discontinuity and multiplicity, Deleuze and Guattari (1987, p. 15) are offering a subterranean critique of the highly influential McCulloch-Pitts model of the neuron and nervous system as functionally binarized, operating either on the basis of a synapse either firing and producing a signal or not; this model is central to the assertion that the human body is organized and cognitively structured in a binary fashion and is thereby amenable to the cybernetic hypothesis. Likewise, when Deleuze and Guattari (1987, p. 16) suggest that information and computer science "still cling to the oldest modes of thought in that they grant all power to a central memory or organ" they are reproaching the work pioneered by early computer scientists like Vannevar Bush, whose proto-cybernetic theory argues that the cognitive processing of the brain is analogous to a digital computer in its hierarchical, syntactic command structure. See: Deleuze and Guattari (1987, pp. 3–25); Pitts and McCulloch (1947); and Bush (2016, pp. 147–157).

3. For a thorough discussion of the reach of cybernetics as an onto-epistemology, see Pias (2015)
4. In their translators introduction to The Three Ecologies, Sutton and Pindar suggest that the persona of Professor Challenger as it is used in the Geology of Morals plateau of A Thousand Plateaus is a parody of the rational scientific subject of modernity. I would add that there is a notably cybernetic twist to this parody. At one point Challenger remarks that he wished to write a programme for a pure computer rather than deliver a lecture, and that his frenetic efforts to posit the Deleuzo-Guattarian philosophy of desiring-production as a universal science are an exercise in "addressing himself to memory only" (Deleuze and Guattari 1987, p. 57), that is, paying homage solely to the arborescent signifier under which any royal science labours. Deleuze and Guattari further parody the philosophy of cybernetics and its literalism by portraying Challenger in quite evocative terms as literally acting out the process of deterritorialization, ultimately being carried off unto the virtual by his line of flight and becoming incorporeal as a result.
5. Statements are a complex, under-utilized component of discourse as it is theorized by Foucault. In short, the efficacy of a given discourse is

premised first upon its unity, which is achieved through the proliferation or rarefaction of certain statements, understood as effective references which are constellated in the articulation of a discourse; making a racist joke publically also makes several statements connected to the culture of racism, stereotypes, tropes, acceptability of racist humour in public. The second criterion of efficacy is the extent to which such discursive unities are taken up and applied, as when a group crafting housing policy who also make racist jokes regularly transmute their sense of humour into concretely affective and institutionalized apparatuses of power within the non-discursive field of discourse.

6. Shannon's showmanship is a distinctive feature of his work on cybernetics, having adopted the term "entropy" and the set of thermodynamic referents it entails to conceptualize his theory of the redundancy of information because it would be both mystifying and alluring to audiences. See Janich (2018, p. 48).
7. Facebook infamously conducted an experiment, widely derided as unethical and immoral, modulating the emotional tone of content represented to certain users in order to investigate the potentially contagious effects of such representations as reflected in the subsequent user interactions with the website, understood through their own emotional tone. See Kramer et al. (2014)
8. The computerization of the world instals a new, rigid form of metering in terms of the rhythmic component of life. Claudio Coletta and Rob Kitchin have considered this state of affairs quite evocatively in terms of the smart city and its conversion of whole urban spaces to this networked, digitally metered form of rhythm. See Guattari (2011b, pp. 109–112) and Coletta and Kitchin (2017).

References

Ashby, W.R. 2015. Homeostasis. In *Cybernetics: The Macy Conferences, 1946–1954—The Complete Transactions*, ed. C. Pias, 593–619. Berlin: Diaphanes.

Bush, V. 2016. Memex Revisited. In *New Media, Old Media: A History and Theory Reader*, ed. W.H.K. Chun and A.W. Fisher, 147–157. New York: Routledge.
Chun, W.H.K. 2011. *Programmed Visions: Software and Memory*. Cambridge: The MIT Press.
Coletta, C., and R. Kitchin. 2017. Algorhythmic Governance: Regulating the 'Heartbeat' of a City Using the Internet of Things. *Big Data and Society* 4 (2): 1–16.
Dean, J. 2009. *Democracy and Other Neoliberal Fantasies: Communicative Capitalism and Left Politics*. Durham: Duke University Press.
Deleuze, G. 1988. *Foucault*. Translated by S. Hand. Minneapolis: University of Minnesota Press.
———. 1990. *The Logic of Sense*. Translated by M. Lester and edited by C. Boundas. New York: Columbia University Press.
———. 1995a. Postscript on Control Societies. In *Negotiations, 1972–1990*, translated by M. Joughin, 177–182. New York: Columbia University Press.
———. 1995b. Control and Becoming. In *Negotiations, 1972–1990*, ed. M. Joughin, 169–176. New York: Columbia University Press.
Deleuze, G., and F. Guattari. 1983. *Anti-Oedipus: Capitalism and Schizophrenia*. Translated by M. Seem. Minneapolis: University of Minnesota Press.
———. 1987. *A Thousand Plateaus: Capitalism and Schizophrenia*. Translated by B. Massumi. Minneapolis: University of Minnesota Press.
———. 1994. *What Is Philosophy?* Translated by H. Tomlinson and G. Burchell. New York: Columbia University Press.
von Foerster, H. 2003. Cybernetics of Epistemology. In *Understanding Understanding: Essays on Cybernetics and Cognition*, 229–246. New York: Springer.
Foucault, M. 1980. Truth and Power. In *Power/Knowledge: Selected Interviews and Other Writings, 1972–1977*, ed. C. Gordon, 109–133. New York: Pantheon.
Guattari, F. 2000. *The Three Ecologies*. Translated by I. Pindar and P. Sutton. London: Continuum.
———. 2006. *Chaosmosis: An Ethico-Aesthetic Paradigm*. Translated by P. Bains and J. Pefanis. Sydney: Power Publications.
———. 2009a. *Microphysics of Power/Micropolitics of Desire' in Soft Subversions: Texts and Interviews, 1977–1985*. Translated by C. Wiener and E. Wittman, 278–290. New York: Semiotexte.

———. 2009b. Plan for the Planet. In *Soft Subversions: Texts and Interviews, 1977–1985*, translated by C. Wiener and E. Wittman, 229–243. New York: Semiotexte.

———. 2011a. Logos or Abstract Machine? In *The Machinic Unconscious: Essays in Schizoanalysis*, translated by T. Adkins, 9–22. New York: Semiotexte.

———. 2011b. The Time of Refrains. In *The Machinic Unconscious: Essays in Schizoanalysis*, translated by T. Adkins, 107–148. New York: Semiotexte.

———. 2016. *Lines of Flight: For Another World of Possibilities*. Translated by A. Goffey. London: Bloomsbury.

Janich, P. 2018. *What Is Information?* Translated by E. Hayot and L. Pao. Minneapolis: University of Minnesota Press.

Kramer, A.D.I., et al. 2014. Experimental Evidence of Massive-Scale Emotional Contagion Through Social Networks. *Proceedings of the National Academy of Sciences of the United States of America* 111 (24): 8788–8790.

Pias, C. 2015. The Age of Cybernetics. In *Cybernetics: The Macy Conferences, 1946–1954—The Complete Transactions*, ed. C. Pias, 11–27. Berlin: Diaphanes.

Pitts, W., and W.S. McCulloch. 1947. How We Know Universals: The Perception of Auditory and Visual Forms. *The Bulletin of Mathematical Biophysics* 9: 127–147.

Shannon, C. 2015. Presentation of a Maze-Solving Machine. In *Cybernetics: The Macy Conferences, 1946–1954—The Complete Transactions*, ed. C. Pias, 474–479. Berlin: Diaphanes.

Stern, D.M. 1993. *Duffless*. Fox Network: *The Simpsons*.

Terranova, T. 2015. Securing the Social: Foucault and Social Networks. In *Foucault and the History of Our Present*, ed. S. Fuggle et al., 111–127. New York: Springer.

7

Inside the Matrix: Matriarchs, Materialisms, and Machinic Being

Janae Sholtz

Conceptions of Technology

Early views of technology assume a sharp divide between the activities of rational human agents and the material world resulting in the almost intractable dualisms between nature and culture.[1] Throughout history, this distinction was rarely challenged, and we can trace its residue through to contemporary philosophers, even those who attempt to restore *techne* to a broader or more fundamental kind of human practice. Heidegger's famous critique of the technization of modernity, in which he laments how thinking has become linked to the manipulation and control of matter. As he conceives it, originally, technology merely meant a kind of

[1] Aristotle, for instance, explains *techne* as "a rational faculty exercised in making something … a productive quality exercised in combination with true reason," while Cicero explains *techne* as the human ability to transform the environment.

J. Sholtz (✉)
Alvernia University, Reading, PA, USA
e-mail: Janae.Sholtz@alvernia.edu

S. S. Das, A. R. Pratihar (eds.), *Technology, Urban Space and the Networked Community*, https://doi.org/10.1007/978-3-030-88809-1_7

knowledge of making or producing,[2] which is paired with *poiesis*, as a kind of bringing-forth guided by *phusis*, provides the answer to our overly technized world—that is, the return of *techne* to its interrelation with nature. His intent is to open up a greater possibility of understanding and communion with the world (and earth) that has been occluded through the association of human knowing with an increasingly curtailed sense of *techne*. But in so doing, he also accepts the Greek distinction between *physis* and *techne*. Now, while this provides an excellent explanation of how technology comes to be doubly removed from nature by its distinction from the broader term for bring-forth or making, *poiesis*, it also exemplifies an ontological dualism that has guided our understanding of the potential of matter itself—to the detriment of our interactions with nature as well as self-understanding. Here we see the early separation of nature and culture that will become contested in works like Donna Haraway's landmark "Cyborg Manifesto."

This view of technology, as the application of knowledge to nature, also becomes the hallmark of our modern scientific understanding, as evidenced in Francis Bacon's publication of *New Atlantis* (1627), where he imagines a perfect society advised by scientists and engineers, who could predict the weather and invent all sorts of fantastical machines that would improve upon nature. Their domination of nature would satisfy all material needs, abolished poverty, and eliminated injustice. This vision helped to inspire the founding of the Royal Society in London in 1662, and institutionalized the belief that science and invention were the engines of progress (Nye 2006, 9). It also provides the paradigmatic view of technology as the outgrowth of a set of ideas or theories developed in the minds of men (geniuses, inventors) and then applied to the material world.

In *Technology Matters: Questions to Live With*, David Nye attempts to lay out an alternative story, one that emphasizes the social aspect of technology, that it comes about "naturally" within contexts of practice, activity, and interaction. On this account, rather than the application of knowledge, theory, or science to matter, technology arises through social evolution and material conditions. The emphasis on material relations

[2] Need citation.

represents a *kind* of materialism, which new materialists will want to say doesn't go far enough, leading to their commitment to analysis focused purely on the materiality of events—the machines, networks that sustain cyber communications, resources that determine form and function, chemical reactions and environmental factors that necessitate or drive ideas and inventions in the first place. Nye traces this alternative story through Lewis Mumford's 1934 work *Technics and Civilization*, in which Mumford proposes technology as a broader network of practices, the sum total of systems of machines and techniques that underlie a civilization. This is an interesting moment in the understanding of technology, in that technology is seen more holistically, no longer as production or making of individual tools or items but as a broad network of interconnected actions and practices. This idea of technology (machines, inventions, techniques) as derived from social relations and reflected in the material conditions of a collective whole is closer to modern theorists such as Foucault or Deleuze, for whom social meaning, subjectivities, and technologies are always collective enunciations resulting from embedded practices

Foucault's description of technology as the working of a collection of practices that produce specific cultural effects aligns with this view and allows for analysis of power within concrete relations (Balsamo 1996, 21). The Foucauldian view is generally viewed as a critical project for identifying how certain taken-for-granted truths are actually constructed. This critical perspective has been used by many feminists to challenge essentialist constructs of gender and sexuality. It is also the basis for discussing how gender does not stand outside of technology but is instead constituted through it—such as De Lauretis' *Technologies of Gender* (1987). In this sense technologies are defined in terms of how the discursive and interpretative are operative in the construction of the material and embodied. In his tribute book to Foucault, Deleuze writes:

> The concrete machines are the two-form assemblages or mechanisms, whereas the abstract machine is the informal diagram. In other words, the machines are social before being technical. Or, rather, there is a human technology which exists before a material technology. No doubt the latter develops its effects in the whole social field; but in order for it to even be

> possible, the tools or material machines have to be chosen first of all by a diagram and taken up by assemblages. … And if the techniques—in the narrow sense of the word—are caught within assemblages, this is because the assemblages themselves, with their techniques, are selected by diagrams. (*Foucault*, 34–35)

This emphasis on social relations as a web of connected knowledges, ideas, and practices (informal diagram), as a kind of human techne that exists before material technologies or material relations, is exactly where new materialists part ways from poststructuralist and the so-called linguistic turn, as well as the implication of human agency (whether collective or individual) implied in Deleuze's resort to "selection." Guattari's description in "On Machines" drives home the commitment to this incorporeal/corporeal dyad with respect to conceptualizing techne:

> The technical object cannot be limited to its materiality. In *techne*, there are ontogenetic elements, elements of the plan, of construction, social relationships which support these technologies, a stock of knowledge, economic relations and a whole series of interfaces onto which the technical object attaches itself. (1995)

Guattari emphasizes the immaterial aspects of technology as a series of interfaces that include collective memory, economic relations, and social relationships. This is not quite the idealism of old, where we are referring to the ideas in the minds of men, but, nevertheless, represents a definite break from object-oriented perspectives through the insistence on the incorporeal universes of knowledge, ideas, and thought. These views of technology seem to suggest some tension, if not an impasse, between the persistent consideration of the immaterial foreground of technology and the most current trends in feminist thought such as new materialist feminisms and posthumanist feminisms, which levy the most stringent critiques of these divisions to date. Are we still caught within the same dualisms between mind and matter, culture and nature? Is this just a repetition of the same historical dualism with which we began or is something else going on here? Where (or what) is technology? Is it in matter or memory—or does this question itself address the very nexus of the problem, where we constantly seesaw between these two identifications? What would the eschewal of the social or cultural in favour of a more autopoietic conception of organization of materialities and relationality mean with respect to technology?

There are three lines of feminist criticism that can be derived from the trajectories just outlined. First, the view of technology as the mastery of nature helps explain some of the early radical and ecological feminist discourses surrounding the equation of the feminine with nature and thus the resistance to technology and its masculine emphasis on control and manipulation. Second, the social view of technology softens this dualism by recognizing the thorough imbrication of materiality and social and discursive practices. Yet, these early considerations of the social failed to acknowledge the gendered nature of the choices surrounding the development of technologies. Social constructivist feminists were instrumental in drawing attention to the fact that activities associated with women's productive capacities were often taken over by men when they became profitable or prestigious (we see this happen with midwifery and medicine)[3] or excised from the category altogether and labelled as merely basic, non-technical activities.[4] Historically, "the increasing adoption of the word 'technology,' [was] not simply a measure of the rise of industrialization. It also measure[d] the marginalization of women" (Nye 2006, 32). Engaging Foucault's analysis of power relations, feminists address ways that gender itself is constructed according to discursive and social practices. For new materialist feminists, however, even this social interpretation of technology still retains too much of the division between mind/matter, and they break away from social constructivist feminists, in order to explore the self-organizing nature of matter and reject humanist suppositions of the anthropocentric exceptionalism. New materialists want to expand this vision of who and what agents of interaction are far beyond the workings of human societies, resulting in what Barad calls agential realism—which would imply that *techne* is within rather than separated from materiality as such, complementing what Bennett refers

[3] In thirteenth and fourteenth centuries men took control of ale-making, a female technology, when it became commercialized and profitable (Nye 2006).

[4] This pertains to the "useful technologies" of weaving or pottery, which eventually came to be eliminated from the category altogether.

to as vital materialism. Yet such a swing of the pendulum involves its own set of tensions. This move towards autopoietic materialism leaves little room for the incorporeal or ideational aspects which Deleuze and Guattari, for instance, emphasize as integral elements of machinic being, and we are left to grapple with the effects for understanding our original concept of technology. I suggest that by returning to Deleuze and Guattari's concept of the machinic as a way of understanding technology and its embedded relations, we may retain a place for sense-making and for politics that gets lost in the posthuman materialist ethos, while also nuancing the claims of social constructivists by expanding the vocabulary of relationality.

Feminist Responses to Technology

The interrogation of technology from the perspective of gender is relatively new, developing as it were alongside the evolution of feminist theory itself. Roughly, we can say that there was an initial split in the reception of technology between *pessimism*: those suspicious of its power as a mechanism of control and the longstanding exclusions of women from its realms and *optimism*: those who saw it as a means of liberation (especially newer forms of digital and biotechnologies) because of how they might sever the link between male privilege and technology. In order to gain a deeper understanding of new materialist feminism and posthumanism, as they seek to reframe the radical contingencies of technology and human existence within a broader matrix of univocal life, it is necessary to clarify the various trends within feminist theory related to technology, moving historically from radical feminist and ecofeminist pessimism through cyber and technofeminist optimism.

Radical feminists and ecological feminists of the 1970s and 1980s, because of their association of technology to masculine paradigms of mastery and control (the first line of criticism) tended to have a rather suspicious attitude towards technology, which manifested as technophobia and resistance to technological development. These suspicions were motivated by the, not-unwarranted, worry that male-driven technologies would be used to usurp the already curtailed domains of female power.

This was a particular issue with the rise of reproductive technologies. The result of rejecting technology as masculine was a return to essentialist claims about women's natural roles: reprioritizing motherhood, celebration of the feminine relation to the earth through the revival of eco-goddesses. Unfortunately, these positions only exacerbated the tech/nature divide by further instantiating it in gender itself. Likewise, rejecting technology as patriarchal only served to reinforce certain assumptions about the sciences as a masculine domain, the residues we see still today in verifiable attitudes and statistical data concerning STEM professions (OCED 2006, 2017).

Cyborg Feminism

In her now canonical *Cyborg Manifesto* (1985), Haraway proclaimed, "I'd rather be a cyborg than a goddess," taking aim at the radical and ecofeminists whose attitudes represented a reactionary and essentialist retreat form technological innovations. Therein Haraway expresses a general optimism with regard to technology, while insisting on the need for critical insight and reconstruction of major theoretical and philosophical paradigms. Haraway's cyborg feminist insights included that the categories of gender, race, and even class are fluid and malleable and can be forged through practices of technoscience themselves; that, basically, there is no natural body. We are all techno-bodies already, influenced by informatics, media technologies, biotechnologies, and so on, and to acknowledge this is a step towards empowerment; that the eradication of the distinction between the natural and cultural or technological represents a policed boundary that inhibits women from participation in scientific and technological discovery. It is in this context that Haraway suggests the conceptual personae of the cyborg, which is also a way of addressing past iterations of feminist encounters with technology as somehow against or outside of gender/subjectivity. The constitutive role of narrative, metaphor, and genealogy that discovered in creating these paradigms led her to advocate for new narratives and new metaphors, especially with regard to imagining the role of women, hence her creation of the figure of the cyborg. The strand of new materialist feminism today that emphasizes the need for speculative fictions and imaginative futures not only links up with Deleuze's call for fabulation and invention, but to Haraway's call for reinvention of female images.

In another of her landmark pieces, "Situated Knowledges" (1988), Haraway levies a powerful critique of the nature/culture divide through her critique of the masculinist assumptions within science and technology. Rather than seeing science or tech/nature divide as essential, she undertakes a genealogy of this supposed essential binary. Historically, she recognizes that paradigms of modern scientific knowledge (objective, rational, logical) are socially constructed on the idea of a modest observer (always assumed male) through which the grand claims of scientific inviolability rest. One of her most well-recognized accomplishments is to expose the ideological fiction of science as detached and objective and further to have exposed the gendered nature of this institution. She argued that science has operated according to a masculinist ideology of mastering and exploitation, propped up by the gendered imagery of nature/culture divide. Understanding science and technology as integrally connected to processes of the production and disciplining of bodies (informatics of domination), rather than neutral or purely objective, is a step towards taking up these technologies for oneself. In other words, if gender, race, class can be seen as both constituted by and constitutive of technoscience, this means that they can be transformed.

Haraway claims cyborgs represent the real hybrid condition of humanity, as always biological and technological. The cyborg is a transgressor of boundaries and embraces technological integration, thus aids in the deconstruction of the paradigm of the "natural female body."[5] Yet, as integrated techno-bio subjects, they do not participate in the myth of the utopic disembodied subjectivity of cyberfeminism. Rather the cyborg is supposed to rupture the boundaries between human and machine in order to proliferate meanings and potential subjectivities. Cyborgs resist pure identities and categorizations and, thus, disrupt gendered binaries, allowing for greater freedom of expression and practices for women (ideally). Haraway does insist that this is a particularly female figure:

[5] Which was never natural but filled with cultural assumptions, interests, and imperatives that serve a particular prevailing social order (see Balsamo 1996, 25–27).

"women have always been cyborg"—in that women have always had to navigate these divisions and possess an inherent awareness of the derisory dualisms which have constituted their phenomenological constraints and socio-political limits. Although few would dispute the idea that Haraway's anti-essentialist feminist subjectivity paves the way for more contemporary versions of feminism which are post gender, posthumanist, and neomaterialist, Haraway's cyborg feminism is not immune to criticism.

Retrospectively, one may detect an accelerationist attitude that actually confirms or at least affirms capitalist and neoliberal paradigms. In *Technologies of the Gendered Body: Reading Cyborg Women* (1996), Anne Balsamo argues that this vision of the cyborg actually re-instantiates liberal subjectivity, due to its emphasis on autopoeticism, autonomy, technological prometheism, and, that Haraway's postmodernist claims to self-construction fail to account for differing material conditions. The female body becomes a matter of technological simulacra—a set of discursive practices and signs that can be manipulated—now by the subject herself. The worry is that "technology transforms the body into nothing more than discourse" (Balsamo 1996, 28). More philosophically perplexing is that the cyborg ends up being synonymous to women, which begs the question, if women have always been cyborg, how does this designation actually address their subordinate social position.

Cyberfeminism

Haraway's cyborg feminism was definitely an important touchstone for those feminists who were enthusiastically wading into the fray of new technologies and theorizing their potentials for women. Cyberfeminism, coined in 1994 by Sadie Plant, describes the work of feminists interested in theorizing, critiquing, and exploring the Internet, cyberspace, and new-media technologies. In their view, the move to postindustrial technological information societies (a digital age) brought with it more flexible networks which could break down traditional roles, identities, localizations. This transformation of technology was seen as a transformation of the traditional masculine nature of technology towards the feminine, privileging what were typically held to be the feminine soft skills of communication over more masculine ones of manual production. Displaying said characteristic optimism, Plant hailed cybernetics as providing instant availability of knowledge, the ubiquity of artificial intelligence, and the near total interconnectivity of the planet.

Emblematized Plant's jubilant proclamation, "only the destruction of the subject will suffice" (Plant, 327), cyberfeminism participated in the promethean spirit associated with transhumanism's desire to utilize cybertechnology to transcend the body altogether. Grounding their reasoning in the recognition that female embodiment has always been particularly inhibitive, cybertechnology was hailed as the ability of women to transcend the biological body—to transcend the categories of woman altogether by participating in a genderless, virtual world where they, women, could be anything they wanted to be. Thus, virtual digital space was met with full-fledged enthusiasm from cyberfeminists, who believed that moving beyond physical divisions meant the ability for women's emancipation: the ability to choose lifestyles (through virtual identities), reinvent the body (through artificial devices, enhancements, and prosthetics), even forego gender roles altogether (biotechnologies related to reproductions and hormonal therapies related to transitioning between sexes for instance).

Cyberfeminists identified several other aspects of cybertechnology as particularly good for feminism, such as how the complexification of public space would up-end many of the traditional assumptions of the public/private divide, how the emergence of networks and contacts which need no central command would evade structures of command and control which predominantly supported the domination of male hegemony and how the introduction of self-organizing systems represents a subversion of patriarchy, where "the goods get together," so to speak, and eradicate the illusion of necessary hierarchy and top-down control (Plant, 328). With respect to technology, cyberfeminism reflects an instance of unadulterated, and perhaps uncritical, optimism,[6] which was later tempered by the recognition that technologies, even virtual ones, don't exist in a vacuum, and that those disembodied subjectivities, first, still exist within a set of social relations to which they must return; second, that the

[6] See "100 Anti-Theses" at the first Cyberfeminist International Meeting in 1997.

very virtual worlds themselves are constructed through the prejudices and values of its users—which happen to be predominantly male; and third, that the idea of a utopic free space fails to account for the influence of capitalism, commercialism, and branding. All of these concerns have, in effect, come to fruition in real time, exemplified by things like Gamergate, which reveals the toxic hypermasculinity and the language of violence rife within online gaming communities, the proliferation of the alt-right, which catapulted a certain kind of violent and fascist rhetoric to the forefront of American politics, and the phenomena of doxing, which has bled over into the policing of academics and intellectual production that dares to call out misogyny and racism.

Another criticism of early cyberfeminism is that their belief in an independent, virtual world which develops language and practices irrespective of the social or material milieu transcends subjectivities, and can therefore remake reality reflected in a tacit endorsement of technological determinism and that regardless of how positively cyberfeminists viewed technology, this view embodies a kind of fatalism, that technology inescapably determines the meaning of reality.[7] For many, the assumption of technological determinism was justified through Marx's powerful analysis of how the mechanization of society reflected the iron law of inevitable historical development, yet others identified problems with this kind of determinism, claiming that mere awareness of particular tools or machines does not automatically force a society to adopt them or to keep them (Nye 2006, 20).

Technofeminism

In *Technofeminism* (2004), Judy Wajcman introduces a new path for addressing the technoscience/feminist relation, and, along the way, levies a critique of Haraway's early work, and cyberfeminism in general, as too quickly heralding the promise of technology to "change" gender relations and move beyond them. She claims that the cyborg is easily subsumed into the pervasive social paradigm and power dynamics that exist, and

[7] Like Marx's interpretation of mechanization of society as part of an iron law of inevitable historical development.

that the sleek hybridity of the machinic cyborg actually comes to signify the hypermasculine worker of manufacturing capitalism (2004, 98), even while the material conditions of actual women reveal predominantly gendered trends—higher demands for service workers with conventional feminine qualities (2004, 98), along with the fact that feminine work/jobs, even technological ones, are seen to be of a lower, less economically valued and socially important class. Wajcman goes as far as the suggestion that this euphoric vision of cyborg society is the vision of a privileged global elite who has access to expensive technoscientific tools for the construction of new identities and reflects an American fascination for technological progress (Wajcman 2004, 99). When Haraway actually speaks about the plight of real women, exploitation of workers globally, or African American oppression, her strategy takes on a more socialist-feminist hue, which actually highlights structures of oppression and structural domination (2004, 99). There seems to be a tension between a kind of "fantastic vangardism" that relies on imagined subjectivities brought about through technology's innovations (the cyborg solution) and the social realities and material conditions of real women and changing social relations, such as the fact that technoscience is becoming increasingly subject to processes of commodification and capitalist accumulation (Wajcman 2004, 90).

Essentially Wajcman's insight is that the material-semiotic approach that Haraway promises gives way to a kind of semiotic play that loses sight of the way that technological change tends to be under the rule of constraint, confusion, and surveillance of actual individuals whose bodies and lives are at stake: "Haraway loses a sense of how feminists could act to change, or at least redirect technologies, rather than just reconfiguring them in writing" (2004, 101). Whereas proponents of Haraway's cyborg-feminism argue that the cyborg reasserts the material body into discussions of technology (Balsamo 1996, 32), Wajcman claims: not enough! The biological body may be present, but the tantalizing promise of discursive transformation is too great, and the emphasis on the material conditions that bodies undergo to faint. What this suggests is that there is more work to be done in moving from theoretical and deconstructive practices to building an actual movement, one that provides guidelines for a practical emancipatory politics. Lacking is a coherent framework that allows us to engage technical change as integral to the renegotiation of gender power relations (Wajcman 2004, 103) and material conditions of existences and located bodies.

Wajcman offers technofeminism as an alternative perspective that builds on the insights of cyborg feminism but grounds it firmly in a thoroughgoing materialist approach to the social studies of technology (Wajcman 2004, 103). Technofeminism returns to ethnography to critique social interests and better understand how technology and gender are mutually constitutive.

Thus, Wajcman's insights provide a more substantial platform for actual political engagement, including the need to transform gender power relations, to change the nature of technological work itself, perhaps by reintegrating work and the personal, to assess where in technology femininity (such as standard versions of mother, girl, woman) is rejected, to address women's and men's concrete socio-technological practices and how these practices are sanctioned (or not), to question how technology is socially acquired and, importantly for me in the context of levying a transnational Deleuzian-inspired critique, who has access to it, and to address the reality of the predominance of men in the design process and that technology designed for women often reflect male prerogatives.[8] The fact that technological products (seen as advancements) for women still seem to revolve around lessening domestic load or time conservation, while products for men are geared towards imaginative projection, entertainment, and enhancing creative interaction with the world underscores this distinction. Technofeminism directs us to the ways that various subjects are situated in relation to technologies, advocating that women need to be seen as more than mere consumers or responders to technology.

All of this background prepares our understanding of the current trends in feminism with relation to technology, capitalism and globalization, social relations, and the environment. In particular, this history provides insight into evolution towards new materialist and posthumanist

[8] Carr's (2018) article "Love, Consent, and Arousal: Deterritorialising Virtual Sex" addresses this issue in relation to the virtual world of cybersex. She argues that for all the imaginative potential, unless women are involved in the design this new technology could very easily fall into a replicatory model of masculine desire.

feminisms. New materialist feminists want to interrogate what exactly is this fleshiness or how are we materially related to other beings, human, inhuman, organic, and inorganic? These questions lead to two major paradigm shifts: the rejection of anthropocentricism (in nature and technology) and the rejection of human exceptionalism. The thread that links these past discussions of feminism and technology with these two movements is a desire to reject the early cyber feminist celebration of disembodiment and also the promethean undertones of cyborg feminism—both of which seem to miss the fleshiness and embeddedness of technology within social relations and, more ontologically, a deeply materialist relationality with the environment and other nonhuman entities and forces. The question for new materialist feminists, is first, are these movements fleshy enough? The answer seems to be no. The further evolution in new materialist thought relates to their core rejection of anthropomorphism on all levels, which puts new materialist at odds with much of the cyborg and technofeminist project focused as they are on socioeconomic and semiotic factors for human societies.[9] My question is what does de-emphasizing the human actually do for feminism and for our understanding of technology and does it go too far?

New Materialist and Posthumanist Feminisms

In new materialist and posthumanist feminists, there are two complimentary ideas touted in relation to technology: (1) Technology becomes imbricated in relational material matrixes that have some level of agential power of their own, rather than seen as merely a tool for human agencies and interests. (2) Technologies have to be considered in light, not just of their impact on human beings, but in terms of how they affect the world of mutual and co-constitutive relations, technology as impersonal milieu. Braidotti, for instance, sees the need to re-theorize the technological apparatus as our new milieu, not as an appendage or prosthetic, but as

[9] Berg (2019) recently advocated a reconsideration of Haraway's cyborg within the recent "turn" in feminist theory towards new materialism, to forge a conversation with agential realism. On her view, the cyborg addresses materiality in a different way than new materialism, from the point of view of the material-semiotic, a concept which includes political critique.

that within which we are inherently and incessantly immersed—experienced as *medianaturecultural.* Said prerogatives suggest fruitful methodologies for widening the framework of what technology touches or effects and the kinds of analyses that we can or should do. Clearly, they must be contextualized within their larger project, and it is by doing so that we may begin to interrogate whether new materialist accounts of technology can really deliver what they promise.

The founding move in new materialist thought is the rejection of social constructivist methods and the critique of what are deemed to be discourse or linguistically oriented understandings of subject formation, and instead to situate subjects within material milieus of differential becoming which highlight the imbrication of bodies (organic and inorganic) and their mutual or symbiotic transformations as not belonging to a particular meaning-making agential entity. The idea is that a politics of representation based on discursive construction cannot grapple with materiality. New materialism purports to address the "neglect of … material phenomena and processes" in cultural and critical theory a field that "privileges language, discourse, culture and values" over the material (Coole and Frost 2010, 3). There is a move away from ideas of causation related to intentional or agential subjects as well, which complicates any discussion of technology, given the longstanding association of *techne* with either the human or at the very least a kind of making which has its cause in a maker rather than itself.

Accompanying this critique of subject/mental-oriented causation is a concomitant reconceptualization of matter as dynamic, and self-organizing. Rather than finding the cause of material transformation within some agent outside of matter, new materialists are wed to the notion of autopoietic materiality—matter has its own agency and vitality. New materialists are interested in "the emergent properties of matter, the movements, forces and processes peculiar to matter and biology" (Frost 2011, 70) beyond human intentionality. This focus supposedly allows feminists to conceive bodies and materiality that resist modes of power and agential manipulations. "New materialists push feminists to decenter human intentionality and design in the conceptualization of the relationship between nature and culture" (Frost 2011, 77). Biology and materiality have as much of an influence on culture as culture has on materiality and embodiment, such that there can be no linear reading of nature/culture, or vice versa. Barad uses the term, "agential intra-action" to capture this vision of biology and culture as co-emergent, and her appropriation of entanglement from quantum theory suggests a relational ontology within which objects always exist within dynamic assemblages (see Barad 2003, 2007). This emphasis on relational ontology suggests that all "entities" are inherently open processes which co-exist in fluid, relationally open milieus beyond subject and object divisions, and theoretically demands a shift towards non-hierarchical and thoroughly egalitarian ethics. Barad's "agential realism" is implicitly egalitarian with respect to the powers of matter—it is the world's performativity of itself that matters, and Braidotti argues explicitly for what she calls zoe-egalitarianism, wherein all organic life, the molecular, chemical reactions, viruses, and biological cellular organisms, are placed on the same level of ethical value. Overall, there is giddy heralding of the human being being dethroned, as Coole and Frost describe it:

> The human species, and the qualities of self-reflection, self-awareness, and rationality traditionally used to distinguish it from the rest of nature, may now seem little more than contingent and provisional forms or processes within a broader evolutionary or cosmic productivity. (Coole and Frost, 20)

This decentring of the human is a fundamental priority which is shared with close theoretical counterpart, posthumanism. These two movements have come to dominate contemporary ethical and theoretical discussions, so much so that they provide the framework for thinking that any ethically conscious, ecology and environmentally aware thinker is supposed to aspire to these days. New materialism aligns with posthumanism in that it decidedly is anti-anthropocentric, stressing that there should be no privileging of human agency, values, or intention. Recognizing that there is a spectrum within posthumanist thought, the general prerogative seems to the undermining of traditional boundaries between the human, animal and technological and rejection of humanist universalism (see Braidotti 2013). Francesca Ferrando (2019) identifies its critical indices as post-anthropocentric, posthumanist, and post-dualist, and considers it a movement towards acknowledging onto-epistemological priorities of openness, alterity, and plurality. Moving away from the (Western European) perspective of human rationalist exceptionalism makes room for acknowledging the meaningful contributions of all beings and the redefinition of the human as an ongoing interactive process, which Braidotti advocates as moving beyond anti-humanism to explore new possibilities of the human (2013, 37). Although Ferrando insists that posthumanists decry techno-enchantment and techno-utopianism associated with transhumanism as "based on anthropocentric and technocentric premises" (2019, 38), it is significant that Katherine Hayles coined the term almost twenty years ago; *How We Became Posthuman: Virtual Bodies in Cybernetics, Literature and Informatics* (1999) reflects the cyberfeminist optimism in moving beyond embodiment. I would argue that the new materialist version of posthumanism is more closely aligned with the celebration of the inhuman that one finds in the work of Gilles Deleuze and engages the philosophical implications of relational ontology which posits a fundamental level of ontogenetic forces and molecular becoming that precedes all form/substance.

Braidotti and Grosz both interpret Deleuze's univocity of being, characterized by molecular and inhuman forces which traverse the human or provide the generative motor of becoming within molar forms, as indicative of a materialist monism that supports the ethical paradigm of nonhierarchical, onto-egalitarianism, as well as indicating a nonhuman turn in philosophy (see Roffe and Stark 2015; Braidotti 2016b). Grosz even claimed that the inhuman is the primary focus of Deleuze's philosophy (2015). This emphasis makes its way into the accounts of new materialism *and* Deleuze studies which eschew discussions of the subject (and the significance of human thought or concept formation) and skew towards subjects of bio-diversity and the environment and inorganic or nonhuman assemblages. It is also evident conceptually in the way that new materialists have appropriated concepts of agency and meaning/mattering as specifically nonhuman phenomena (see Barad 2003, 2007).

In *Critical Posthumanism and Planetary Futures* (2016), Braidotti confirms that the explosion of work on inhuman, nonhuman, and the posthuman triggered by anti-humanism and anti-anthropocentrism, critiques of species hierarchy, and the emphasis on ecological justice have been, by large, influenced by Deleuze and Guattari's univocal, monist account of being.[10] Yet, Braidotti contends that *her* posthumanism recuperates a place for subjectivity (i.e. human), and she calls reactions against posthumanism that associate it with poststructuralist prerogatives towards the death of the subject "knee-jerk," which actually suggests to me that she recognizes that new materialism has problems dealing with the particularities of human subjects. She says that monism doesn't eliminate differences but dismantles hierarchical structures that have coded them, allowing us to see that the world and humans are materially embedded subjects in process of circulating nomadically within webs of relations. These connections produce assemblages that she defines as transindividual forms of subjectivity (Braidotti 2016, 27) that include nonhuman factors and technological mediation. For her, the posthuman necessitates a methodology that combines monistic neo-materialism with feminist politics of location (embodiment and lived experience—emphasizing the situated nature of knowledge). Braidotti wants a nature-culture continuum which stresses embodied and embrained immanence involving negotiations with bio-genesis, neurosciences, environmental sciences, race, gender ethnicity, and disability studies (2016, 19).

Yet it is difficult not to see her proposal of an affirmative composition of transversal subjectivities that is expanded beyond the human, coupled with her ethical imperative of *zoe-centred egalitarianism,* as belying this "emphasis" on human subjectivity and undercutting the potpourri of ethical considerations for which she claims such a critical posthumanism can and does account. Further, although Braidotti may have recognized this theoretical weakness, it is not clear that such nuance has made its way through to the manner in which new materialism takes its equalizing gaze to its subjects of study—biological, chemical, animal—where the cultural and linguistic have all but been banished by dint of new materialism's founding gesture. The problem is not in assuming that the posthuman participates in the anti-humanism of poststructuralist deconstruction of the subject, but that the appropriation of agency and meaning-making as kinds of equilateral material practices negates

[10] While it is true that Deleuze and Guattari reject the idea of the autonomous, humanist subject, they do not reject the human or shift their focus away from subjects. Deleuze absolutely maintains a concept of subjectivity, albeit radically transformed from the independent subject/ego to how collective enunciations form subjectivities that are then occupied by individuals (taken up).

meaningful distinction between autopoietic mattering, or meaning as performative arrangements, and the kinds of thought processes that make up human practices and complexities. In "Matter in the Shadows: Feminist New Materialism and the Practices of Colonialism," molecular biologist, Deboleena Roy, and organismal biologist, Banu Subramanjam, worry about "the curiously disembodied nature of new materialist distribution of interest" that makes it possible to focus on bacteria, birds, lizards, and so on "without having to refer to the embodied materialities effected through their contact with political systems of power such as gender, class, race, or colonialism" (2016, 34). Similarly, Stephanie Clare, in "On the Politics of New Materialism" (2016), suggests that the faith in nonhuman agency and inorganic vitality is overly optimistic, that breakdown of species boundaries is attended with a certain idealism, which fails to account for the incorporation of animality within biopower. She suggests that the celebration of the inhuman democracy of subjects does not attend to different positions (economic, race, class) that affect these relations. Although mostly everything that I read of new materialism at least suggests that social issues of race, gender, class, and so on can be incorporated, somehow, into their theoretical framework, the emphatic focus on onto-egalitarianism and deprivileging of the anthropomorphic belies these claims.

Regardless of gestures to shore up the theoretical problems that arise from such thoroughgoing anti-anthropocentricism and anti-human exceptionalism, the theoretical positions which hold sway in new materialism impede our ability to have these discussions. Haraway's advocacy of multispecies feminism in *Staying with the Trouble* comes to mind, where she advocates for living-with critters—spiders, ants, pigeons—while arresting human procreation: "make kin not babies." Sophie Turner has an excellent critique of this position in a short piece entitled "Cthulhu Plays No Part of Me," where she says this move to multispecies feminism is characterized "by a barely disavowed willingness to see whole cities and cultures wiped from the planet for the sake of a form of thriving among 'companion species' involving relatively few of *us*."[11] Haraway's book is well titled and I want to say "yes, let's 'stay with the trouble!'" But the

[11] https://www.viewpointmag.com/2017/05/08/cthulhu-plays-no-role-for-me/.

emphasis seems misplaced if it means moving away from the human, because, by and large, we are the trouble.

Frost says that "what is at stake in thinking in terms of complexity, interdependence, and ecology broadly construed is *epistemological and political humility* in the face of the organic and inorganic world" (Frost 2011, 79). But is this enough to constitute a politics? We have to address the human desires and mindsets that cause the despicable decisions related to the defacement of our environments and which feed the corporate machine, and we need to reconsider the meanings that we give things and the particular methods that we use to select for particular assemblages (of course, taking into consideration our material relations and the thoroughly external nature of collective enunciations that form subjects). And what has happened to the concept of technology itself? It all but falls out of the discussion of what enter into assemblages—the biological and the social/culture seem to be accounted for, but the technological—it seems to get subsumed in the idea of "social relations." Of course, fostering greater sensitivity to our milieus, our materiality, our connections, and relations to other living things is important, but in this move to the posthuman and celebration of ontological monism, we seem to have thrown the baby out with the bathwater and watered down our ability to say anything meaningful about actual pragmatic conditions of human interaction—to speak about meaning, to make meanings is a particularly human endeavour that I want to reintroduce into the equation.

I'm not the first to make such a critique. In "Deleuze and New Materialism: Naturalism, Norms and Ethics" (2017), Keith Ansel-Pearson acknowledges that Deleuze is important for new materialism but that he also "bequeaths a complex legacy to contemporary thought about the human" (2017, 88) and he worries that the posthumanist appropriations are overlooking the clear focus on the ethical task of human emancipation therein. Ansel-Pearson argues that, though Deleuze's materialism does entail that there is no cosmic exceptionalism, this does not mean that there aren't assemblages of greater complexity or that there aren't distinctive human potentials to be ethical and make decisions. Pearson turns to Deleuze's early essay "Instincts and Institutions" to show that Deleuze understands the human species as ethically and politically distinguished by its propensity for invention of both institutions and norms.

What Pearson takes issue with, as do I, is the key implication in new materialism, that "there is no privileging of human bodies or even human capacities for agency" (2017, 90). This makes little sense of Deleuze's emphasis on subjectivity, agency, the human capacity for invention and selection, or ethics. Pearson identifies this humanist impetus in Deleuze's work with his reading of Spinoza, but we can find it immediately in the concept closest to materialist appropriations—the machinic *assemblage*, the purported motor of his materialist vision of the plane of immanence.

Recuperating the Machinic

The idea of the machinic is developed through Deleuze and Guattari's collaborations, notably in *Anti-Oedipus*, where it is used to describe the way that desire operates as a series of cuts and/or interruptions in the perpetual flow of desire *(AO, 36)*; *Kafka*, where it is used to designate a principle of external functioning and transformation rather than internal signification; and in *A Thousand Plateaus*, where they propose the machinic phylum and describe the ontogenetic operations of materiality as a perpetual flow of matter linked to the concepts of immanence and univocity. It is primarily this version of the machinic, as a descriptor of a relational, self-transformative ontogenetic level of immanent being that new materialists most align with. Vis-à-vis their concepts of the machinic assemblage and the plane of immanence that it conjures, one can posit a vital monism that cuts through the common distinctions between kinds of systems or being. The machinic, rather than being a characteristic of machines or technological objects, is indicative of a principle of autopoiesis which covers organic and inorganic matter alike; Deleuze and Guattari do not maintain a categorical separation between natural entities and manufactured artefacts, a position which corresponds to new materialists' rejection of dualism between nature/culture, while providing an explanation for the genesis of both. Machines, and in particular, abstract machines also provide the cuts and contours by which "objects" and assemblages are carved out of the plane of immanence. In all instances, the machinic indicates a radical openness and potential for transformation and connectivity. Machines plug into other machines, and perhaps most significantly, this view of interaction suggests that objects are significant because of their relationality rather than interiority. Bodies are all conceived on a machinic plane of immanence, and their meaning or function is defined according to what other machinic assemblages they are connected with or are able to connect to.

Although new materialists have to a large extent dropped this language of the machinic, in favour of more vitalist language like vibrant matter, they retain the fundamental features of this Deleuzo-Guattarian ontology: *assemblage, ontogenesis, univocity, transversality*. Yet, these options of Deleuze's relational ontology and assemblage theory also neglect some important aspects or features of assemblages, univocity, and the like, which complicate the new materialist appropriation of Deleuze and Guattari's ontology for their vital materialist picture. This may just be a matter of emphasis, but in my view, it is what gets left out, that provides a necessary bridge between the considerations and critiques outlined above, and, therefore, the path through Deleuze and Guattari, especially with regard to their machinic ontology and its complementary components, machinic assemblages, should be revisited.

On the one hand, the machinic does indicate a univocal level of forces and flows (the aspect that new materialists identify as the monist unity of all beings). This is the machinic phylum: "We will call an assemblage every constellation of singularities and traits deducted from the flow of matter-movement. The assemblages cut the phylum up into distinct, differentiated lineages, at the same time as the machinic phylum cuts across them all" (Thousand Plateaus, 406). The way that Deleuze and Guattari speak of the machinic phylum suggests the element of self-organization that new materialists employ in their own vitalist, agential-realist notions of materiality: "The *machinic phylum* is materiality, natural or artificial, and both simultaneously; it is matter in movement, in flux, in variation, matter as a conveyor of singularities and traits of expression. … This matter flow can only be *followed*" *(409)*. Thus, the machinic phylum does suggest matter as dynamic and filled with potentials (intensive forces) and does also suggest a plane of immanence in which all forces and bodies are essentially connected (ontological univocity). But, on the other hand, this machinic phylum is inseparable from organization, stratifications, and territorializations and provides a *raison d'etre* for how one moves between a vision of becoming as pure variation, and random, heterogeneous relations towards a world of determinations, consistencies, and organization.

To address this second aspect of the machinic, we have to move beyond machinic phylum to the level of abstract machines and incorporeal universes:

> The concrete machines are the two-form assemblages or mechanisms, whereas the abstract machine is the informal diagram. In other words, the machines are social before being technical. Or, rather, there is a human technology, which exists before a material technology. No doubt the latter develops its effects in the whole social field; but in order for it to even be possible, the tools or material machines have to be chosen first of all by a diagram and taken up by assemblages. (Foucault, 34–35)

Although there is an ontogenetic level to machinism, Deleuze refers to diagrammaticism, concept creation, and principles of selection that happen on and within planes of immanence as equally indicative of machinic being—in other words, a plane of immanence has to be constructed and that construction is both a matter of material variable relations and incorporeal, conceptual relations. Unfortunately, assemblage has become popularized to the point of losing the nuance of Deleuze and Guattari's intent. As Buchanan (2020) has observed, Delanda, for instance, views all of Deleuze's concepts pertaining to multiplicity, machinic being, heterogeneity, and assemblage solely in terms of heterogeneous arrangements of matter. Delanda's view collapses the distinction between abstract machines and concrete machines—the latter's abstractness merely relating to the kind of materiality (unformed) with which it deals. It is this Deleuze that is really taken up by new materialists—a purely materialist Deleuze.

As Guattari's three ecologies environmental, social, and mental suggest, there are multiple levels and strata that must be considered and the point of the abstract machine it is able to connect these different strata of materiality and semiotics, bodies and knowledge—this is transversality—not just cross-species or intra-relations among different material bodies, molecular, chemical, organic, inorganic as materialists seem to suggest. There are two important factors in this description that have significant implications. First, machinic assemblages are not merely combinations of material elements. This may seem overly simple, but, even if this is tacitly understood, it isn't addressed sufficiently in new materialist uptakes of the concept. Machines are neither solely material nor purely incorporeal. Abstract machines, which are immanent aspects of machinic assemblages, link material and semiotic elements, moving between entirely heterogeneous series, not just different levels of materiality. These heterogeneous series can include physical objects, events, and material flows, but also include signs, utterances, and forms of expression. In fact, abstract machines are inextricably composed of forms of content (materiality) and forms of expression (which I am interpreting as a kind of sense-making). In other words, the machinic does not suggest a bare life or materiality. They are assemblages that include multiple ecologies—biological, environmental, and mental, and we could add historical social, and linguistic, and although new materialists are often given credit for a more complex view of assemblage, when one drills down to its usage, it often amounts to more or less complex arrangements of material. The "assemblages" that come to be highlighted are usually examples of material relations, and the "transversal" cross over between species and between inorganic and organic things.

Second, Deleuze's focus on sense-making (how relations get expressed) is an important element of Deleuze's account of machinic materialist being. It may be helpful to remember what Deleuze finds significant about Stoic logic in *Logic of Sense*. The Stoics, although thorough materialists, posited sayables as incorporeals, as effects of material interactions and bodies. What is expressed, these incorporeal attributes, nonetheless have a kind of quasi-effect, in the sense that they express ways of being within an assemblage which have an effect: carving out locations, roles, and so on. It is by reinvigorating the virtual potentials within actualized relations, expressing them, that abstract machines can break away from order-words/commands that constrain these relations—offering new pathways for thinking and being. This is not just a matter of reconfiguration of bodies and materials. New expressions unleash incorporeal transformations as well as having a recursive effect that potentially setup new corporeal spaces and configurations.

What is often missing in new materialist accounts is the discussion of meaning and the production of sense, and how this incorporeal surface can be a quasi-cause which re-infuses matters and unlocks different potential modes of connection. Karen Barad, one of the most notable new materialists to directly theorize meaning, conceptualizes the materiality of meaning by incorporating the ontological insights of quantum physicist, Niels Bohr. Barad endorses Bohr's insights that, because the apparatus used to measure or "see" the phenomena is an eliminable part of the phenomena itself, concepts are just these fully material-specific physical arrangements, but extends his thought to all material arrangements. Barad extends the ontological implications by claiming that phenomena are not the mere result of laboratory exercises engineered by human subject and apparatuses don't just mean the technological machine used to conduct an experiment, thus the need to posit an observer who arranges an experiment is eliminated; matter as always in various stages of self-organization and intra-action takes on the agential roll. Her formulation of discursive practices as material configurations upon which human beings have no particular priority or purchase seems to be the bedrock for new materialist discussions of meaning and signification (see Selberg and Hinton 2016).

Yet, Deleuze clearly reserves a particular space and role for human thought and concept creation via abstract machines. Deleuze and Guattari stress that the machinic includes principles of selection and that abstract machines can themselves become constraining machines of capture that mould the way of seeing and being in the world, and thus are sense-making machines rather than just significations of meanings or material configurations that have meaning. Machinic assemblages can be either constrictive or liberatory, vis-à-vis the principles of selection and the "procedures which deterritorialize its elements," a key way that the machinic exceeds the new materialist focus on the autopoietic self-organization of matter.

Given that the role of social liberation is fundamental in Deleuze and Guattari's work, this elision of sense-making and the de-emphasis on the human are potentially detrimental to these causes. In moving from explanations of material machinic systems to systems of oppression and capture, you have to move through social relations which are composed as an assemblage of forms of content and forms of expression. Technological enslavement does involve reducing what is thinkable through binarization and taking for granted a human-centred world, but the criticism of the anthropocentricism is not reason to forego the recognition that it is human attitudes and mindsets that must change along with configurations of material relations.

Technocapitalism

Though new materialists often decry the inability of prior feminism (social constructivist) to account for a robust and autonomous materiality, a common criticism of new materialism is that its preference for analyses of a universalized material is too abstract, that is unable to account for differential social relations and considerations of race and class. Several feminist critics have argued that new materialists are limited by their universalizing conceptualization of matter and practically exclusive focus on the priority of a politics of life and its affirmation. Julie Torrant (2014) takes issue with Braidotti's emphasis of sustainability where the measure of sustainability is the equality of matter/life (see Braidotti 2010). The effect of this focus on "sustainability" is the abandonment of the project of women's emancipation and the effects are "deeply problematic when considered in relation to the material conditions of working-class families" (2014: np). In the name of the universal priority of materiality, new materialism bypasses the ensemble of social relations and historical conditions that produce social contradictions in capitalism and presents contradictions as transhistorical and existential conditions of life as such. Axelle Karera levies an even more pointed assessment with respect to race, claiming that posthumanist and new materialist affirmation of bio-interconnectivity "sidestep an engaged account of social antagonisms, and more specifically those enacted along racial lines. Instead, these are smoothed over and displaced in the name of an ethics of futurity grounded on a deeply naturalized variation of relationality" (2019, 43) and that they are "unequipped to account for the ongoing and brutal victimization of black people" (2019, 47).

This to me seems to be a good reason to retain Deleuze and Guattari's concept of the machinic assemblage as method that attends to both striations (social relations) and material desires and affective relations (investments of desire). Take, for instance, Lazzarato's analysis of the machine of capitalism, and its specifically contemporary iteration, technocapitalism. Utilizing Deleuze and Guattari's model, Lazzarato insists that investment in the technocapitalist machines happens both at the levels of the social identities, class and racial divisions, roles and functions that are necessary to sustain the assemblage, and pre-individual, pre-cognitive desires or inhuman forces, what Lazzarato calls the "invisible side of capitalism," claiming that technocapitalism is "a series of devices for machinic enslavement" vis-à-vis the affective, corporeal milieus as well as "social subjection" vis-à-vis the social roles and identities that come to be defined and constrained through capitalism. The ability of providing an analytic tool for assessing and transforming both is the additive critical value of Deleuze and Guattari's concept of the machinic.

So, how does the technocapitalist machine constitute racially, socially, and globally differentiated women, and are our theoretical models complicit in supporting machinic enslavement? Torrant offers a scathing critique of new materialism in the respect, one that echoes Wajcman's pointed criticism of cyborg feminism as a technofantasy available only to a privileged global elite. She argues that "new materialism" is a *ruling class movement* in cultural theory in general and in feminist theory in particular aimed not only at ideologically and pragmatically adjusting exploited workers to the exigencies of capitalism, but to shore up the class privileges of a small ruling class minority. Essentially, her argument is that it is a kind of neoliberal privilege to speak about how nice we should be to spiders or that we should scale back our human desires so as to lead balanced lives with biological others when we are destroying economies and lives of global others in the name of capitalism. Some women cannot scale back their desires any more—they are already robbed of the privilege of expressing desires, needs, or emotional demands.

Roy and Subramanjam (2016) advocate for an account of materiality that is simultaneously an account of power and privilege, suggesting that feminist science studies scholars interested in new materialism might want to take a look at the critical modes of analysis that are being conducted in postcolonial and transnational studies, claiming that this allows a more accurate account of axes of influence. Some of the prerogatives that are introduced via these studies are situatedness, articulation of local points of interest within larger networks, contact zones of empire, the destabilization of the central to become local, and the global effects of capitalism which seem to prey upon particularly vulnerable bodies of women, children, and the poor in regions of economic depression. Speaking directly against new materialist feminists, they observe that "It's not just the contemporary body and their biology that we must think about, but also the historical constructions and disciplining of Indian bodies, as gendered and raced." To exemplify this difference, they refer to the chemical devastation of methyl isocyanate on various populations in India, showing that this is not just an issue about how the materiality of the situation is detrimental—how the molecular properties interact and constrain the environment, but insist that it matters where and on whom these chemicals are being dumped—but about who it is okay to dump toxic chemical on, whose bodies to taint, are worthless. Transnational feminism has developed a more nuanced account of the gendered effects of capitalism on a global scale. What often gets left out of the discussion of technology is that "many" women, globally speaking, encounter technology as something that they are producing for a global economy in which they are minimal participants, not as something that they will use or manipulate but something that represents danger, violence, and the precariousness of their livelihoods.

Donna Haraway says that "It matters what thoughts think thought, what stories tell stories" (2016, 35), and that story must be no longer one of human exceptionalism (2016, 39). For Haraway, we compose communities through speculative fictions and storytelling and our community should be eco-bio-zoologically centred, for which she conceives the idea of composting as a new form of being-with: "I compost my soul in this hot pile. The worms are not human; their undulating bodies ingest and reach, their feces fertilizes worlds" (Haraway 2016, 35). This is to create new material matrixes that trace the patterns of all the complex nonlinear couplings that occur around us and live according to their cadences, rather than tromping around in our big human boots wreaking havoc as we go. It is hard to argue with Haraway that it is imperative for us to do something to make our environments less toxic and more liveable—and that this necessarily includes considering how we can co-exist with other living beings, or that we should develop practices of caring that look at entire milieus or assemblages. We are thoughtless if we do not hold open a space for the other; but the other could be a pigeon or a spore. Which leads me back to the question, what is the trouble? It's us. It's capitalism. It's our choices. It's our desires.

It cannot be just about limiting our technological or environmental footprint, or learning to exist in more sustainable relations to the environment—these don't make sense in light of the ways that individuals are differentially constructed by the state, capitalism, and the access to technologies. It is clear that this kind of emphasis on unequal distribution of roles and functions is important to any analysis of material relations. That may solve part of the problem, but I wonder—how does that change people's attitudes and minds and desires? Assessing these kinds requires an analysis of the material forces and the routinization of desires that are involved in our reactions to certain subjects and activities, as well as in the construction of certain relations of dominance. We must couple this analysis with the articulation of desire, as well as the proliferation of new desire—the level of machinic enslavement—and ask ourselves, how are desires and affects routed to justify these predicaments and how do we find the desire to break with these machinic codings of our desire?

We have to think about new matrixes from which new stories and new ways of acting, thinking, and being arise in terms of both material relations and the desires and modes of affective organization that guide us. We have to think in terms of new machinic matrixes, not just material matrixes. As we saw, machinic enslavement happens through the capture of affect and desire. Thus, *the responsibility we have is to address our own pitifully routinized desires—what and where are we investing desire—what prejudices and judgments do this lead to?* Deleuze and Guattari say that "what belongs to all requirements of liberation: the force of the unconscious itself, the investment by desire of the social field, the disinvestment of repressive structures" (AO, 69). What do we prioritize? What does it mean to be a subject in the twenty-first century? If it is the case that we are experiencing the "homogenization of perceptual experience at an unprecedented level," how do we disrupt this steady flow of affective anaesthesia. We have to find ways to create new matrixes, that do not abide by the same operations of power and privilege, that seek to include the voices and the experiences of the marginalized, that tend to their desires and understandings of the world, that offer opportunities for creative expression and modes of connection that empower and make visible their lives. In terms of political mobilization, what is needed is to engender spaces for the concerns and desires of those at the margins, not just to create awareness but to take precedence—an ethical imperative that establishes priorities rather than eschewing them.

References

Ansel-Pearson, Keith. 2017. Deleuze and New Materialism: Naturalism, Norms, and Ethics. In *The New Politics of Materialism: History, Philosophy*, ed. Sarah Ellenzweig and John H. Zammito. New York: Routledge.

Balsamo, Anne. 1996. *Technologies of the Gendered Body: Reading Cyborg Women*. Durham: Duke UP.

Barad, Karen. 2003. Posthumanist Performativity: Toward an Understanding of How Matter Comes to Matter. *SIGNS: Journal of Women in Culture and Society* 28 (3): 801–831.

———. 2007. *Meeting the Universe Halfway: Quantum Physics and the Entanglement of Matter and Meaning*. Durham, NC: Duke University Press.

Berg, Anne-Jorunn. 2019. The Cyborg Its Friends and Feminist Materiality. In *Discussing New Materialism*, ed. U. Kissmann and J. van Looh. Wiesbaden: Springer.

Braidotti, Rosi. 2010. The Politics of 'Life Itself' and New Ways of Dying. In *New Materialisms: Ontology, Agency, and Politics*, ed. Diana Coole and Samantha Frost, 201–218. Durham and London: Duke University Press.

———. 2013. *The Posthuman*. Polity Press.

———. 2016. *Critical Posthumanism and Planetary Futures*. Edited by D. Banerji and M.R. Paranjape. Springer.

———. 2016b. Anthropos Redux: A Defense of Monism in the Anthropocene Epoch. *Frame* 29 (2): 29–46.

Buchanan, I. 2020. *Assemblage Theory and Method*. London: Bloomsbury.

Carr, Cheri L. 2018. Love, Consent, and Arousal: Deterritorialising Virtual Sex. *Deleuze and Guattari Studies* 12 (4): 597–611.

Clare, Stephanie. 2016. On the Politics of 'New Materialism. In *Mattering: Feminism, Science, and Materialism*, ed. Victoria Pitts-Taylor, 58–72. New York: New York University Press.

Coole, Diana, and Samantha Frost. 2010. *New Materialisms Ontology, Agency, and Politics*. Durham: Duke University Press.

De Lauretis, Teresa. 1987. *Technologies of Gender: Essays on Film, Theory and Fiction*. Indianapolis: Indiana University Press.

Ferrando, Francesca. 2019. *Philosophical Posthumanism*. London: Bloomsbury Press.

Frost, Samantha. 2011. Implications of New Materialism for Feminist Epistemology. In *Feminist Epistemology and Philosophy of Science: Power in Knowledge*, ed. H.E. Grasswick, 69–83. Springer.

Guattari, Félix. 1995. On Machines. *Complexity, JPVA* 6: 8–12. http://archtech.arch.ntua.gr/forum/post2006interaction/on_machines.htm.

Haraway, Donna. 1985. Manifesto for Cyborgs: Science, Technology, and Socialist Feminism in the 1980s. *Socialist Review* 80: 65–108.

———. 1988. Situated Knowledges: The Science Question in Feminism and the Privilege of Partial Perspective. *Feminist Studies* 14 (3): 575–599.

———. 2016. *Staying with the Trouble*. Durham: Duke University Press.

Hayles, Katherine. 1999. *How We Became Posthuman: Virtual Bodies in Cybernetics, Literature and Informatics*. University of Chicago Press.

Karera, Axelle. 2019. Blackness and the Pitfalls of Anthropocene Ethics. *Critical Philosophy of Race* 7 (1): 32–56.

Lazzarato, Mauritzio. 2006. The Machine. In *Transversal Texts*, translated by M. O'Neill. http://eipcp.net/transversal/1106/lazzarato/en. Accessed December 16, 2020.

Nye, David. 2006. *Technology Matters: Questions to Live With*. Cambridge, MA: MIT Press.

OECD. 2006. *Evolution of Student Interest in Science and Technology Studies. Policy Report*. www.oecd.org/dataoecd/16/30/36645825.pdf. Accessed December 12, 2020.

———. 2017. *The Pursuit of Gender Equality: An Uphill Battle*. Paris: OECD Publishing. https://doi.org/10.1787/9789264281318-en.

Roffe, J., and H. Stark 2015, "Deleuze and the Nonhuman Turn: An Interview with Elizabeth Grosz", in ed. J. Roffe and H. Stark, *Deleuze and the Non/Human*. Palgrave Macmillan.

Roy, Deboleena, and Banu Subramanjam. 2016. Matter in the Shadows: Feminist New Materialism and the Practices of Colonialism. In *Mattering: Feminism, Science, and Materialism*, ed. Victoria Pitts-Taylor, 23–42. New York: New York University Press.

Selberg, K., and P. Hinton. 2016. Introduction: The Possibilities of Feminist Quantum Theory. *Rhizomes: Cultural Studies in Emerging Knowledge* 30: 1–15. https://doi.org/10.20415/rhiz/030.e03.

Torrant, Julie. 2014. It's Time to Give Up Liberal Bourgeois Theories, Including New Materialist Feminism, and Take Up Historical Materialist Feminism for the 21st Century. *The Red Critique* 15: np. http://www.redcritique.org/WinterSpring2014/historicalmaterialistfeminismforthe21stcentury.htm. Accessed December 16, 2020.
Wajcman, Judy. 2004. *Technofeminism*. Polity Press.

8

Posthuman Urban Spaces in Dave Eggers' the Circle

Pramod K. Nayar

This chapter examines the construction of posthuman spaces and their geographies. Using Dave Eggers'[1] celebrated novel, *The Circle* (2013), I examine the construction and performance of identity within the crucible of information, technology and surveillance structures, producing a new posthuman subjectivity itself.

[1] Although Eggers does not dwell so much on the biological aspect of the new technologies as much as cyberpunk (including the genre-defining films in the Matrix series) does, it is useful to be conscious of the intersection of biology and biological sciences with informational technology. Eugene Thacker's theorization of biomedia, in particular, is useful here. Thacker defines biomedia as the 'technical recontextualization of biological components and processes' (2004: 1). Corporeality and biological materiality are aligned with an extratechnological moment that is consonant with the 'body itself'. The biological body is 'compiled' through modes of visualization, modelling, data extraction and in silico simulation (13). These technologies prioritize the biological domain as a set of components in interactions with each other via various biomolecular, biochemical and cellular processes (14–15).

P. K. Nayar (✉)
Department of English, University of Hyderabad,
Hyderabad, Telangana, India

S. S. Das, A. R. Pratihar (eds.), *Technology, Urban Space and the Networked Community*, https://doi.org/10.1007/978-3-030-88809-1_8

Eggers' novel has been described as a 'dystopian science fiction novel' (Konstantinou 2013), although, as Roy Sommer notes, it is 'not a book about tomorrow: the dangers the novel illustrates are too familiar to be ignored today' (2017: 55).

I begin with an argument about the performativity of humans in the posturban landscape as an informational performativity. In the second part of the chapter I argue that information undertakes a specific performance of the spaces occupied by the characters of *The Circle*, and resonates with a certain neoliberal Gothic mode, as theorized by Linnie Blake.

Informational and Posthuman Performativity

Urban spaces are increasingly negotiated as a congeries of the SAT-NAV and GPS, the screen, the camera and the human body in ways that reconfigure those spaces. The city therefore embodies both a 'splintering urbanism' and a techno-corporeal urbanism. I take it as a truism Debra Shaw's argument that 'the power relations of posturbanism can be read through the new cartographies of bodily distribution that it produces' (2017: 10, emphasis in original).

Graham and Marvin (2001) identify a 'splintering urbanism' in contemporary cities. The 'unbundling', as they term it, of infrastructure so that the larger public is now increasingly marginalized in favour of specialized packages for the IT or business sector marks the 'fragmentation of the social and material fabric of cities' (ibid.). The creation of SEZs, cybercities, techno-parks and financial districts in metropolises, all heavily surveilled and monitored, is the splintering urbanism of Graham and Marvin. And yet this explanation and theorization of urbanism seem inadequate. I propose that any urbanism today demands, in the main, corporeal disbursement in the sense of Shaw's 'new cartographies'. That is, urbanism is activated, motivated and driven by new configurations of embodiment and bodies—human and even nonhuman. This embodiment is within information, first and foremost. As I shall argue, informational performativity is posthuman performativity in the contemporary urbanscape.

If performativity is the employment of speech acts and language to define and assign identities, in the age of information capitalism and the digital, the predominant mode of enacting identities is via information: compiling, presenting, sharing and forwarding. Informational performativity is the enactment of one's identity through the collating and sharing of personal data in online worlds, but also responding to others' data flows. While at the surface level this corresponds to a form of impression management, Eggers' novel takes it far beyond that: to document every waking moment of one's life, from eating to working, is to perform an acceptable identity. For Mae Holland, acceptance, recognition and even self-acknowledgement are contingent upon sharing information. Further, her responses to others' information or the initially unnerving but eventually welcome step of being always visible constitute her. Informational performativity is akin to, but has a more embodied and manifest 'effect' to the 'informatized human' theorized by Katherine Hayles (How We Became Posthuman). In Hayles' pithy formulation, 'the posthuman view privileges informational pattern over material instantiation' (1999: 3). The point is: this informational pattern is no longer distinct or distinguishable from the material body of Mae Holland, her everyday actions and her emotional make-up because she is responding to the informational pattern coming to her, even as she seeks to fit into the pattern by offering personal data of her own. That is, in the posthuman scenario theorized by Hayles and fictionalized by Eggers, her material instantiation is a consequence and source of her informatization, or what I am calling informational performativity. Informational performativity (re) produces or (re)constructs the 'somatically legible subject' (Nash 2011) as the somato-informatically legible subject.

In Dave Eggers' *The Circle*, Eamon Bailey, one of the founders of the company, Circle, points to this specific feature:

> Imagine any city with this kind of coverage. Who would commit a crime knowing they might be watched any time, anywhere? My friends in the FBI feel this would cut crime rates down by 70, 80 percent in any city where we have real and meaningful saturation.

Bailey is not speaking of technologies of surveillance alone, although that is integral to what goes on in contemporary urbanscapes, but rather of how bodies constitute the fabric of a (securitized) city, along with buildings, sensors, networks and other lifeforms. The 'who' in the above excerpt is the human who brings alive a certain kind of geography, supposedly safe, secure and crime-free. The larger point of such a city is that all humans have to be surveilled in order to deter some from engaging in crime. In other words, the city's safety can only be ensured by placing all bodies inhabiting it, traversing through it, participating in it, within the scopic regime of surveillance mechanisms.

Posthuman cities are spaces of advanced scopic regimes where bodies are enmeshed into and yet vivify the very mechanisms developed to restrain them. Further, I propose that in Bailey's words—and the entire edifice and community of the Circle—we can see what Karen Barad would term posthuman performativity (2003).

To begin with, Bailey's words here and elsewhere and indeed of the account of lives led by the community speak of a material discursivity. In Barad's terms,

> materiality is discursive (i.e., material phenomena are inseparable from the apparatuses of bodily production: matter emerges out of and includes as part of its being the ongoing reconfiguring of boundaries), just as discursive practices are always already material (i.e., they are ongoing material (re)configurings of the world). (822)

'Discursive practices', writes Barad, 'produce material bodies' (808).

I will illustrate Barad's argument regarding posthuman performativity and its relevance as a heuristic to read urban spaces through the account of the elevator in Eggers.

Eggers describes an elevator experience that Mae has on her first tour of the Circle:

> They entered an elevator of glass, tinted faintly orange. Lights flickered on and Mae saw her name appear on the walls, along with her high school yearbook photo. WELCOME MAE HOLLAND. A sound, something like a gasp, left Mae's throat. She hadn't seen that photo in years, and had

> been happy for its absence. This must have been Annie's doing, assaulting her with it again. The picture was indeed Mae—her wide mouth, her thin lips, her olive skin, her black hair, but in this photo, more so than in life, her high cheekbones gave her a look of severity her brown eyes not smiling, only small and cold, ready for war. Since the photo—she was eighteen then, angry and unsure—Mae had gained much-needed weight, her face had softened and curves appeared, curves that brought the attention of men of myriad ages and motives. She'd tried, since high school, to be more open, more accepting, and seeing it here, this document of a long-ago era when she assumed the worst of the world, rattled her.

Renata, who is showing her around, says:

> "Yeah, everything's on sensors". "The elevator reads your ID, and then says hello. Annie gave us that photo. You guys must be tight if she's got high school pictures of you. Anyway, hope you don't mind. We do that for visitors, mostly. They're usually impressed."

Eggers writes:

> As the elevator rose, the day's featured activities appeared on every elevator wall, the images and text traveling from one panel to the next. With each announcement, there was video, photos, animation, music.

First, the sensors construct a certain version of Mae: an old photograph places before her eyes and that of the world a different 'body' that is Mae. What and who Mae is, is the effect of three material-discursive practices: her ID card, her image on the elevator and her living-breathing body occupying space and volume in the elevator. Then, Mae's own visceral and sensory experience of seeing herself telecast is uncanny: it is her and yet not her, since the uncanny is the mixing of temporalities (unpaginated). An old Mae-as-discourse and a current one (in the flesh) are merged in the space of the elevator. The performativity that produces Mae's identity, then, in Barad's words, 'incorporates important material and discursive, social and scientific, human and nonhuman, and natural and cultural factors' (808). In the Circle, the ID card and the photograph are both material and discursive. Socially and culturally, the company believes that 'all that happens must be known' (unpaginated). The human—Mae—and the nonhuman—the elevator—are in constant intra-action. Then, the 'natural' body of Mae and the discursive one embodied as her photograph (which itself has two forms: an archival footage somewhere and its reproduction on the elevator walls).

Second, Renata explains how the picture was sourced: from a schoolmate of Mae's. But what she says later is the key: 'visitors are impressed', presumably by the effect of seeing older photographs of themselves telecast on the screens and walls. The emotional response of Mae and the visitors, which determines the affective experience of this specific topos, is determined by both discursive and material practices. A childhood or past is brought alive into the present. The elevator sensors record her identity today and bring up her past. The materiality of the space asserts its effect on the individuals through this technologically (nonhuman) mediated visual and aural (both image and the background announcements) discourse.

These together reconfigure Mae's present and the immediate world she occupies—the Circle. A visual text from her past, presenting her then body, arriving in the present when she has a different body, provokes a certain response from her: she does not like it. 'Just when she couldn't stand it anymore, the photo disappeared', writes Eggers. The performativity that is Mae's identity, then, is material-discursive. The company itself speaks of this material-discursive identity and person that emerges as a congeries:

> all of every user's needs and tools, into one pot and invented TruYou—one account, one identity, one password, one payment system, per person. There were no more passwords, no multiple identities. Your devices knew who you were, and your one identity—the TruYou, unbendable and unmaskable—was the person paying, signing up, responding, viewing and reviewing, seeing and being seen. You had to use your real name, and this was tied to your credit cards, your bank, and thus paying for anything was simple. One button for the rest of your life online.

In the posthuman urbanisms of the contemporary this performativity aligns the corporeal body, its projections and representations and discourses that reconfigure the spaces occupied and experienced by the human individual embedded in them. Rather than a splintering urbanism, one sees a posthuman-mediated urbanism here.

Eggers' novel critiques this form of posthuman urbanism and technospaces when he shows how relentless the pressure to contribute to this performance of the community and the self can be.

That information retrieval and storage, before being put to use by capitalism and the corporations is now a truism, and Eggers puts it explicitly:

> the actual buying habits of actual people were now eminently mappable and measurable, and the marketing to those actual people could be done with surgical precision.

Mae discovers that she has to constantly post and repost responses to others' posts, and contribute her own, to be a 'good' member of the Circle. There is a PartiRank (Participation Rank) based on how active she is. This PartiRank, Gina tells her, is taken to be a Popularity Rank, 'but it's not really that'. It is

> just an algorithm-generated number that takes into account all your activity in the Inner Circle.

It takes into account zings, exterior followers of your intra-company zings, comments on your zings, your comments on others' zings, your comments on other Circlers' profiles, your photos posted, attendance at Circle events, comments and photos posted about those events—basically it collects and celebrates all you do here. The most active Circlers are ranked highest of course.

Mae experiences very early, what this means: unstoppable information flows, most of which is shared personal information for public consumption.

> The third-screen feed dropped forty new Inner Circle messages every few minutes, fifteen or so Outer Circle posts and zings, and Mae used every available moment of downtime to quickly scroll through…

When she does not join newsfeeds about particular activities, she is questioned about it, just as her colleagues and bosses are upset by the fact that she did not post anything on social media about her kayaking.

Community participation in the posthuman urbanscape of *The Circle* is measured in terms of information flows. Information is material in the social spaces of the company, and material experiences must have a representational component that is shared with the entire community. Information matters in the dual sense that it is seen as the building blocks of social relations—Mae is given a lecture on the shared etymological roots of 'communication' and 'community'—data, in short, matters.

This is informational performativity that is at once material and discursive where 'Mae' as matter does not preexist her discourses/representations, her relationships with(in) the Circle are the effects of the intra-action between her body and its data circulating in the networks, and finally, the Mae-as-subject and Mae-as-object dichotomy is performatively enacted through the material-discursive.

It is precisely this level of communication and information sharing—Circle's slogans are 'Privacy is Theft' and 'Sharing is Caring'—that her ex-partner, Mercer, an environmentalist and artist, objects to. Mercer refers to the new social media tools and their functions as the construction of neediness in his quarrel with Mae:

> the tools you guys create actually manufacture unnaturally extreme social needs. No one needs the level of contact you're purveying. It improves nothing. It's not nourishing.

Identity as informational performativity that is embodied in the Circle's insistence on Mae's sharing every aspect of her life on the company's social media generates 'neediness', according to Mercer, and this is not a healthy development. In Mercer's language, which may be construed as Eggers' voice:

> You no longer pick up on basic human communication clues. You're at a table with three humans, all of whom are looking at you and trying to talk to you, and you're staring at a screen, searching for strangers in Dubai.

Mae buys into the Circle's ideology, and becomes the company's face, agreeing to be filmed round the clock. From this moment, her rise in the company is nothing short of meteoric, indicating that mobility in the profession also comes from losing all sense of privacy. In Roy Sommer's words,

> The greatest risk to individual freedom, Eggers reminds us, is no longer Big Brother but the naive (and greedy) user who unwittingly pays for the "free" services of web browsers and social networks with the gold and oil of the digital age: personal data. (2017: 53)

One performs personal data for the world to consume: in this performance is the identity, and social mobility, that one craves and eventually internalizes. Mae becomes 'the benign, friendly face of it all'.

This is the new informational performativity where older forms of communication have been irretrievably lost and replaced by the social media. In Shaw's terms, 'spatial practice constructs a correspondence between information space and the lived space of the body' (6), as socio-technical assemblies or 'congeries' (Nayar 2014: 64). Mercer's argument, one could say, is directed at the posthuman congeries who emerges in these urban spaces: always monitored, connected, enmeshed in informational flows so that the person is not separable from her avatars.

Information, Performative Cartography and the Neoliberal Gothic

In Eggers' novel, the motto of the Circle is 'all that happens must be visible'. Even before she is made aware of this motto, Mae encounters the embodied practice of the slogan during her first encounter with the company's offices:

> She pushed open the heavy door. The front hall was as long as a parade, as tall as a cathedral. There were offices everywhere above, four floors. High on either side, every wall made of glass. Briefly dizzy, she looked down-

ward, and in the immaculate glossy floor, she saw her own face reflected, looking worried.

She 'entered an elevator of glass, tinted faintly orange'. The actual offices 'were fronted by floor-to-ceiling glass, the occupants visible within'. The cafeteria is no different: 'the diners ate at nine different levels, all of the floors and walls glass. At first glance, it looked like a hundred people were eating in mid-air'.

The office is turned inside out so that the boundaries between the inside and outside are rendered blurry and transparent at the same time. The hyper visible space of the Circle offices is the architectural equivalent of the motto 'all that happens must be visible'. The activities of the workers in the space of the Circle's office are a sophisticated system of monitoring and controlling workers' bodies. By keeping every cubicle and office space behind glass walls, this visual standardization of workers' bodies is also a form of behaviour regulation. That is, the workers in Circle are monitored so that their actions and behaviour become subconsciously conformist with the company's policies, exactly in the same way that Mae discovers that she has to post, repost and respond to the many activities, events and personal messages appearing on her screens. While the Circle's employee tells Mae, 'We're not automatons. This isn't a sweatshop', the disciplinary regime constructed out of visibility ensures that the workers in that space regulate themselves precisely because they are visible. (Larger questions of labour and gender in such spaces need to be addressed, but that is not the subject of this essay.)

The information-based performance of cartography has a clear capitalist-imperial angle as well. On her tour of the offices, Mae encounters young men examining a simulation:

> They met a pair of women working on a submersible exploration craft that would make the Marianas Trench mysterious no more. 'They'll map it like Manhattan', Annie said. … They stopped at a table where a trio of young men were looking at a screen … displaying 3-D drawings of a new kind of low-cost housing, to be easily adopted throughout the developing world. (unpaginated)

This is emblematic, writes John Masterson, of the Circle's 'empirical and expansionist zeal, in the form of mapping enigmatic depths, with the supposed altruism of development discourses' (2016: 732). Information enables the mapping of and control over all spaces, including distant ones. Mae recognizes this, but sees it as something positive; she sees the company's work

> as a show of strength, and as a demonstration that with the will and ingenuity and economic wherewithal of the Circle, no earthly question would remain unanswered.

This is in fact unbeatable power, driven by technology and information.

The 'circle', in other words, will envelop the earth. There are no geographical, territorial or cultural boundaries and everyone can be a consumer. As Eggers puts it:

> the actual buying habits of actual people were now eminently mapable and measurable, and the marketing to those actual people could be done with surgical precision. (unpaginated)

And Eamon Bailey says in the novel: 'There will be very few populated areas that we won't be able to access from the screens in our hands'.

There will also be no resistance possible. Eggers writes:

> And there was a wonderful thing that tended to happen, something that felt like poetic justice: every time someone started shouting about the supposed monopoly of the Circle, or the Circle's unfair monetization of the personal data of its users, or some other paranoid and demonstrably false claim, soon enough it was revealed that that person was a criminal or deviant of the highest order. … And it made sense. Who but a fringe character would try to impede the unimpeachable improvement of the world?

Information, controlled and total, will produce this 'improvement'—or what the company terms 'Completion'.

In his incisive reading of the novel, Masterson writes:

> the rhetoric of surgical precision is transposed from a discourse of "heavy security" (military strikes, drone warfare, and collateral damage caused by the technological advances of supposedly postmodern conflict) to one that, on the surface at least, appears much softer: that of late, if not too late capitalism. Here targets on a screen are consumers rather than enemy combatants, with the chilling suggestion that they are neutralized in similarly devastating ways. (732)

Linnie Blake (2019) argues a case for the Neoliberal Gothic: 'Neoliberal Gothic … repeatedly returns us to ways in which we have been made monstrous by the workings of capital in a world that espouses free trade but in which we are far from free' (61, emphasis in original). Examining contemporary TV series and popular Hollywood films, Blake writes:

> the vampires of the neoliberal age not only embody and enact the parasitic nature of contemporary laissez-faire economics but enable us to observe how readily human power-brokers have abandoned their humanity in their ongoing pursuit of wealth. (62)

For Blake, the linkage between the government, biomedicine and capitalism is the crucial factor in the new—and monstrous—forms of global hegemony. In Blake's view, the neoliberal Gothic is effected through the marking of several populations as unworthy of living. The conquest of homespace, workspace, the emotions and bodies of 'subjects' and their predictive functioning through networks and information flows render those like Mae's older parents vulnerable.

The sense of being haunted by the processes, expectations and the constant pressure of having to 'fit in' is something Mae comes to experience early on in the job. This is a posthuman scenario in which the subordination of the human/person to mechanical, software and other processes does render them into the kind of zombie-like labour Blake sees in contemporary neoliberal Gothic. Blake—also following Marx's famous account of the vampire-like processes of capitalism—writes of the zombie as a 'mutation of the neoliberal self into a corporatised brand identity struggling to find a vestigial humanity in the rubble of a broken world'

(67). The zombie figure has been seen as a response to the dehumanizing nature of contemporary global capitalism (Lauro and Embry 2008).

The brand identity of belonging, of being 'in' the circle, of sharing one's life through data networks is an onerous weight.

Mae checked her clock; she'd been on the Orlando query for eight minutes, far surpassing the new guideline per query, which was 2.5.

Mae discovers that she has to speed up her response time, and each second is clocked and contributes to her ranking. Even with her best efforts, she falls behind and the absence of speedy enough participation results in responses such as this:

> Sorry to be Ms. Sensitive, but after I invited you to choose my professional network, you didn't ask me to join your professional network, and though I know I'm just a nobody in Orlando, I felt like I had to tell you that it made me feel devalued.

Being valued and bestowing value are mediated processes in the Circle. The numbers that clock up with each act of posting and reposting ensure a sense of belonging:

> So many of her watchers, they reminded her, working at desk jobs, too, and because she continued to do this work, voluntarily and with evident joy, they saw her as a role model and inspiration. And this felt good. This felt truly valuable to Mae. The customers made her better. And serving them while transparent made her far better. She expected this. She was apprised by Stewart that when thousands, or even millions, are watching, you perform your best self. You are cheerier, more positive, more polite, more generous, more inquisitive. But he had not told her of the smaller, improving alterations to her behavior.

Now this seems innocuous enough: Mae is being evaluated for her contribution to the social sense of the organization. But the passage that follows immediately implies a conditioning that neoliberal horror stories consist of:

> The first time the camera redirected her actions was when she went to the kitchen for something to eat. The image on her wrist showed the interior

> of the refrigerator as she scanned for a snack. Normally, she would have grabbed a chilled brownie, but seeing the image of her hand reaching for it, and seeing what everyone else would be seeing, she pulled back. She closed the fridge, and from the bowl on the counter, she selected a packet of almonds, and left the kitchen. Later that day, a headache appeared—caused, she thought, by eating less chocolate than usual. She reached into her bag, where she kept a few single-serving aspirin packets, but again, on her screen, she saw what everyone was seeing. She saw a hand searching her bag, clawing, and instantly she felt desperate and wretched, like some kind of pill-popping addict.

She did without. Every day she'd done without things she didn't want to want. Things she didn't need. She'd given up soda, energy drinks, processed foods.

This is behavioural programming, and one that critiques of neoliberal forms of governance identify. It homogenizes the workforce, and their attitudes. The Circle is posthuman civilization that is at once highly industrialized and dehumanizing.

Yet, the novel does not present the Circle as a disciplinary institution, in the Foucauldian sense. Indeed, the novel is an exploration of 'coercive corporate culture rather than a disciplining institution', as Frida Beckman puts it (2020: 539). It is important to see how performative cartography, where receiving and sending information is the performance of the spaces one occupies is not, contra Beckman, about the 'difficulties of accommodating for the dissolution of identities and institutions in the society of control' (540). Instead, it enacts the performance of a neoliberal subjectivity and brings it close to the genre of the 'neoliberal Gothic'.

First, there is the substitution of militaristic language and the discourse of securitization with the language of belonging and identities. Bailey observes of Mercer who, in their view, is resisting information-based integration: '[Mercer is a] deeply depressed and isolated young man', who could not 'survive in a world like this, a world moving toward communion and unity' (unpaginated).

Here is Mae's set of thoughts that imply connection, security, order and systems—all driven by information and algorithms. Mae's thoughts are all about performing an informatized cartography:

> [Mae had] had enough of the chaos of her family. … On campus there was no friction. She didn't need to explain herself, or the future of the world, to the Circlers, who implicitly under- stood her and the planet and the way it had to be and soon would be.

Increasingly, she found it difficult to be off-campus anyway. There were homeless people, and there were the attendant and assaulting smells, and there were machines that didn't work, and floors and seats that had not been cleaned, and there was, everywhere, the chaos of an orderless world. The Circle was helping to improve it but in the meantime, it was increasingly troubling to be amid the madness outside the gates of the Circle.

Walking through San Francisco, or Oakland, or San Jose, or any city, really, seemed more and more like a Third World experience, with unnecessary filth, and unnecessary strife and unnecessary errors and inefficiencies—on any city block, a thousand problems correctible through simple enough algorithms and the application of available technology and willing members of the digital community.

Technologically ordered/controlled worlds are, in Mae's vision now, as the passage shows, better than the chaos of the family, the street and the city itself. One notes in the passage, the movement outward, from family to the 'Third World' itself, metaphorically encompassing in rhetoric the movement and aspirations of the company itself.

That said, Eggers also suggests that there are dark Spaces in such a neoliberal Gothic's posthumanist geographies.

Mercer asks in the novel:

> Did you ever think that perhaps our minds are delicately calibrated between the known and the unknown? … You people are creating a world of ever-present daylight, and I think it will burn us alive. There will be no time to reflect, to sleep, to cool.

Mercer is referring to spaces beyond the map, beyond the informational conquest of companies like the Circle. These are the dark spaces of techno-neoliberal Gothic sites.

Debra Shaw speaks of the 'the ideology of urban space as always already "regenerated"; cleansed, devoid of "dark space" and, more importantly, wholly visible' (118). The term 'dark space', from Anthony Vidler (1992), is deemed to be a space that hides objects of fear and danger that may trouble the imagination of the people and are inimical to their happiness. In Shaw's reading of posturbanism and posthuman spaces:

> The back lighting which provides for full visibility in the architectural arrangements of the panopticon is designed not only to banish the dark space where improper practices can be hidden but to bring enlightenment to both the occupants of the cells and the administering bureaucracy. (125)

For disciplinary regimes, notes Shaw, 'the pairing of transparency and obscurity' is the key in controlling the (human, living) bodies in that space (127). More importantly, for Shaw

> It [dark space] is a prime site for the emergence of imperceptible politics, that is, that the inevitable materialisation of dark space alongside the architectural provision of light space creates unthought opportunities for the performance of insurgent posthumanism. The problem with dark space, of course, is that it manifestly prohibits vision; the sine qua non of utopian political projects. (127)

Mercer seeks such dark spaces:

> I'm moving north, to the densest and most uninteresting forest I can find. I know that your cameras are mapping out these areas as they have mapped the Amazon, Antarctica, the Sahara, etc. But at least I'll have a head start. And when the cam- eras come, I'll keep going north.

When Mae seeks to find Kalden (who turns out to be Ty, one of the three men who founded Circle), in the company's premises, she is unable to. The irony is not lost on her when she thinks:

> She checked CircleSearch again; she'd looked for him a hundred times this way, with no success. But she had a right to know where he was. To at least know where he was, who he was … She could know, instantly, the tem-

> perature in Jakarta, but she couldn't find one man on a campus like this? … Mae cursed the not-knowing, and knew she needed someone who could be known. Who could be located.

In a context where the newest monitoring device is announced with pride (because 'transparency leads to peace of mind', as Eamon Bailey puts it), Mae is unable to locate one man.

Yet, such dark spaces must not be allowed to exist, even if these are within the human/person. We see this idea emerging towards the very end of the novel.

When Annie, her friend, goes into a coma, Mae visits her in hospital. The thoughts that go through Mae's head are the last lines of the novel:

> Another burst of color appeared on the screen monitoring the workings of Annie's mind. Mae reached out to touch her forehead, marveling at the distance this flesh put between them. What was going on in that head of hers? It was exasperating, really, Mae thought, not knowing. It was an affront, a deprivation, to herself and to the world. She would bring this up with Stenton and Bailey, with the Gang of 40, at the earliest opportunity. They needed to talk about Annie, the thoughts she was thinking. Why shouldn't they know them? The world deserved nothing less and would not wait.

Mae's is nothing less than a posthuman vision: of being able to penetrate Annie's thoughts (rendering Annie a cybernetic organism in her coma?). The company's aim, and now her own, is a world where everyone could know each other truly and wholly, without secrets, without shame and without the need for permission to see or to know, without the selfish hoarding of life (unpaginated).

But Mae's meditation about Annie also illustrates the point about posthuman spaces that Shaw raises: of spaces that are still, despite all the efforts to map them, dark.

Writing of biopunk dystopias, Lars Schmeink observes the genre's emphasis on genetic and bio-engineering, posthuman subjectivity, the human body as the site of conflicting ideologies of capitalism and resistance, monstrous bodies, among others (Schmeink 2014). I have elsewhere argued that the 'posthuman Gothic is not one where the urban space has been taken over by the undead: these are spaces wherein humans who are not recognizably humans abound' (2016: 79).

Conclusion: The Subjects in Posthuman Urbanscapes

Debra Shaw summarizes the critical posthumanism position succinctly:

> In an irony which seems to confirm the argument for technogenesis, it is the emerging products of a global culture invested in describing life in such a way that it can be represented as strings of code, patented, manipulated and interfaced with consumer electronics. (49)

While Eggers does not speak in terms of code, he does refer throughout to the manifestation of those codes: the algorithms, the processes, the social media that run on codes. These are the constituents of the new subjectivity of the posthuman condition.

Mae's subjectivity, like those of others in the Circle, hinges on information flows and her ready immersion in them. Shaw writes:

> The forms of subjectivisation effected by posturban space emerge out of the way that these same discourses have responded to the kinds of immaterial labour and disembodied sociality promoted by the ascendancy of information space. (126)

The serial updating and posting, responses and counter-responses are Mae. The 'disembodied sociality' of the online worlds that Mae takes to, and transforms as a consequence, her material life into an online series for consumption by whoever is logged in or following her produces the posthuman subject that is Mae-on-camera. Posthuman capitalism alters her cognitive responses and her emotional ones.

The media, writes Zizi Papacharissi, creates the 'affective feedback loops that generate and reproduce affective patterns of relating to others that are further reproduced as affect—that is, intensity that has not yet been cognitively processed as feeling, emotion, or though' (2015: 22–3). Mae is precisely this: fitting into affective patterns of feeling without clear processing. Each time Mae wishes to eat or do something that has been a habit with her, the affective pattern into which she has been now immersed comes into play. Karen Barad's arguments about the material-discursive and the intra-action that is agential appear manifest to an extraordinary degree in Mae Holland's informatized life in Eggers' novel. In the totalitarian informational gaze that is the Circle, only the posthuman subject can survive. That Mercer—who wishes to escape this form of posthuman/informatized subjectivity—dies when seeking to avoid the drones (operated by the Circle as an instantiation of the totalitarian informational gaze) and driving his car off the bridge is Eggers' strongest critique of the new informational economy represented by the Circle.

The affective patterns are modes of using data-feeds from individuals to control them and others, via massive technologies. Resonant with the rise of what Andrea Fumagalli (2012) describes as 'cognitive biocapitalism' (Nash 2011) the Circle attempts to generate knowledge through informatization, even if this is information about personal likes and dislikes over food, sports or clothing. This is then used to serve a system that, as always, makes use of the 'bare vital faculties of human beings' (8). The result is the Mae at the end of the novel: a posthuman subject in posturban space.

Works Cited

Barad, Karen. 2003. Posthumanist Performativity: Toward an Understanding of How Matter Comes to Matter. *Signs* 28 (2): 801–831.

Beckman, Frida. 2020. Control and the Novel: Dave Eggers and Disciplinary Form. *Modern Fiction Studies* 66 (3): 527–546.

Blake, Linnie. 2019. Neoliberal Gothic. In *Twenty-First-Century Gothic: An Edinburgh Companion*, ed. Maisha Wester and Xavier Aldana Reyes, 60–86. Edinburgh: Edinburgh University Press.

Eggers, Dave. 2013. *The Circle*. Alfred A. Knopf. eBook.

Fumagalli, Andrea. 2012. Twenty Theses on Contemporary Capitalism (Cognitive Biocapitalism). Trans. Sabrina Ovan, *Angelaki* 16 (3): 7–17.

Graham, Stephen, and Simon Marvin. 2001. *Splintering Urbanism: Networked Infrastructures, Technological Mobilities and the Urban Condition*. London and New York: Routledge.

Hayles, N. 1999. *Katherine. How We Became Posthuman: Virtual Bodies in Cybernetics, Literature, and Informatics*. Chicago and London: University of Chicago Press.

Konstantinou, Lee. 2013. Dave Eggers Is Worried about America. *The American Prospect*. 13 October 2013. Accessed 29 November 2020. https://prospect.org/culture/books/dave-eggers-worried-america/.

Lauro, Sarah Juliet, and Karen Embry. 2008. A Zombie Manifesto: The Nonhuman Condition in the Era of Advanced Capitalism. *Boundary 2* 35 (1): 85–108.

Masterson, John. 2016. Floods, Fortresses, and Cabin Fever: Worlding 'Domeland' Security in Dave Eggers's *Zeitoun and The Circle. American Literary History* 28 (4): 721–739.

Nash, Richard. 2011. Joy and Pity: Reading Animal Bodies in Late Eighteenth-Century Culture. *The Eighteenth Century* 52: 47–67.

Nayar, Pramod K. 2014. *Posthumanism*. Cambridge and Malden, MA: Polity.

———. 2016. Appupen's Posthuman Gothic: The Snake and the Lotus. *South Asian Review* 39 (1–2): 70–85.

Papacharissi, Zizi. 2015. *Affective Publics: Sentiment, Technology, and Politics*. Oxford University Press.

Schmeink, Lars. 2014. *Biopunk Dystopias: Genetic Engineering, Society, and Science Fiction*. Liverpool: Liverpool University Press.

Shaw, Debra Benita. 2017. *Posthuman Urbanism: Mapping Bodies in Contemporary City Space*. London and New York: Rowman and Littlefield.

Sommer, Roy. 2017. Beware the Siren Servers: How Techlash Novels Like Dave Eggers's *The Circle and Jarett Kobek's I Hate the Internet* Make the Need for Change Feel Real. *StoryWorlds: A Journal of Narrative Studies* 9 (1–2): 51–70.

Thacker, Eugene. 2004. *Biomedia*. Minneapolis and London: University of Minnesota Press.

Vidler, Anthony. 1992. *The Architectural Uncanny: Essays in the Modern Unhomely*. Cambridge, MA: MIT Press.

9

From Miasma Theory to Digital Ghost Town: Tales of Infrastructure and Social Politics in the Twenty-First-Century Megalopolis

Vinzenz Hediger

The jet-setting son of a rich big-city family finds love in a rural setting governed by stern morality; a mysterious underground figure returns from abroad to start a political movement and fight a cabal of drug dealers threatening to hijack the government; an architect starts receiving messages from a former lover he had left behind in the quarter where he grew up and which he helped raze and transform into valuable property for investors, including his wealthy father-in-law: Three narratives are chosen at random, one from Nigeria (Tope Oshin's 2018 Nollywood comedy "Up North"), one from India (Prithviraj Sukumaran's 2019 Malayalam action thriller "Lucifer") and one from South Korea (Hwang Sok-yong's 2015 novel "At Sunset"), which share neither genre nor medium but two common traits—they are set against the backdrop of urban environments shaped by migration, and in them digital communication devices serve as props which move the narrative forward.

V. Hediger (✉)
Institut für Theater-, Film- und Medienwissenschaft,
Frankfurt am Main, Germany
e-mail: hediger@tfm.uni-frankfurt.de

S. S. Das, A. R. Pratihar (eds.), *Technology, Urban Space and the Networked Community*, https://doi.org/10.1007/978-3-030-88809-1_9

Together these three narratives provide an experimental set-up, if you will, for a comparative analysis of three similar, yet different configurations of narrative, prop and urban space.

What is the use of such an analysis?

My claim is that such configurations of narrative, prop and urban space can contribute to an understanding of what Ravi Sundaram proposes to call the "media urbanity" of contemporary urban settings, and particularly the megacities which are pivotal sites of social and cultural transformation in the twenty-first century. What is significant about the three examples is that the digital communication devices serve as props which are both parts of infrastructure and point to infrastructure as formative of the urban experience.

Contemporary urban environments in their often bewildering complexity pose a problem of intelligibility. In the humanities and social sciences this challenge has been tackled through a variety of approaches, perhaps most productively through ethnographic studies of infrastructure (Leigh Starr 1999) and similar work in economic geography (Engel et al. 2018). To study narratives about and informed by infrastructure in film and literature can complement these approaches, but also add a new perspective. Narrative is the default mode of temporal ordering and memory (Schank 1995). Stories, as Paul Ricoeur has argued, solve the philosophical puzzle of time by giving time an intelligible form (Ricoeur 1990). Narrative is thus not just a way of world making, but of making worlds intelligible. Furthermore, if the content of a medium is always another medium, as McLuhan famously claimed (McLuhan 1964), cinema in particular has a long history of framing and making intelligible other media, from radio to television to video and digital media (Hediger 2010). Digital media devices in cinematic and literary narratives are thus not just props, but media in media, which in and of themselves induce a reflection on media and the mediated condition of the narrative itself and of the world it renders intelligible.

In that sense, just like environmental humanities study the cultural and social aspects of the environmental crisis which are beyond the grasp of the natural sciences (Heise 2017), an analysis of narratives of infrastructure and media urbanity can provide a key to urban life as lived experience, to phrase it in phenomenological term. At the same time, such narratives can provide a key to technological and spatial underpinnings of that experience itself. Ravi Sundaram demonstrates this in a chapter on traffic and accidents in Delhi, which he opens with a discussion of Uda Prakash's short story "Paul Gomra ka Scooter", which is about a newspaper clerk who buys a scooter to facilitate his daily commute but is overwhelmed by the traffic, never learns how to drive and ultimately goes mad (Sundaram 2011, 139).

Ultimately, the question is also one of community and of politics. Urban migration, media urbanity and digital infrastructures induce new types of sociability, but also new types of conflict. Established categories of class, ethnicity, religious identity or culture can only partially account for the multilayered communities in arrival cities. At the same time, and somewhat paradoxically, urban migration reinforces the salience of these categories in conflict: extreme inequality, urban poverty and communal violence are part and parcel of life in the transitional spaces of arrival cities. From an area studies point of view, these new communities provide the final proof that the "container model" of culture, which proposes to study contained territorial communities, is no longer operative. Following Arnika Fuhrmann's (2016) suggestion, we can argue that cinematic (and by implication, literary) narratives provide a key to an understanding of the transformation of the spatiotemporal frameworks that inform social and political agency under the conditions of urban migration. And, as Ravi Sundaram points out in response to Partha Chatterjee's concept of a "political society" and of the politics of mobilizing the "governed", which includes squatting, the regularization of unauthorized colonies and the building of informal infrastructures in urban environments, there is no "politics before media", but only politics in and through media (Sundaram 2011, 21).

In what follows, I first want to discuss the question of media urbanity and the intelligibility of urban space with a view to the question of what Sarah Sharma proposes to call "temporal publics", that is the multilayered temporalities of urban space. I then want to address the question of narrative and props and/as infrastructure, before I discuss the three examples in the essay's experimental set-up.

Megacities, Media Urbanity and the Politics of Temporal Publics

Cities, and particularly the fast-growing megacities in Africa, Asia and the Americas, are the site of the most important social transformation of the twenty-first-century urban migration. According to the UN's World Urbanization Prospect report of 2018, 55% of the world population currently live in cities. That share is expected to rise to 68% by 2050 (UN 2018). Los Angeles, Lagos, Istanbul, Karachi, Delhi, Guangzhou, Shanghai, Beijing, Manila or Seoul are "arrival cities", to use Doug Saunders' term, that is large conurbations which precariously absorb new arrivals from the countryside or other countries, ultimately integrating them in new urban arrangements (Saunders 2011). But cities are also the "factories of the future", to quote economist Richard Baldwin (Baldwin 2016). As manufacturing continues to dwindle as a share of the economy and the importance of information technology and cultural industries grows, megacities emerge as both important centres of production and primary markets, as well as poles of what Fatima Bhutto proposes to call, with not a little irony, the "new world order of cultural production", in which Paris, London, New York and Los Angeles no longer dominate the global supply chains for cultural goods (Bhutto 2019). The growing urbanization goes hand in hand with new forms of urbanity, in particular with what Ravi Sundaram in his study of the transformation of Delhi proposes to call "media urbanity" (Sundaram 2011), a complex, dynamic network of media technology, precarious infrastructure and informal economies, which can be found in similar configurations around the world, with Lagos as another example (Larkin 2008). One can argue, of course, as Shannon Mattern does convincingly in her work on the materiality of urban spaces through history, that cities have always been "smart" and built for intelligence (Mattern 2017). However, the recent turn towards infrastructure in media studies has shown, among other things, that communication technologies add a different layer to the urban experience, not least by creating new spatiotemporal configurations which call established notions of urban space and the dichotomy of "city" and "country" into question (Parks and Starosielski 2015;

Mukherjee 2019). In particular, digital communication and remittance economies bind "country" and "city" together in new ways. If William Cronon, in his re-reading of Turner's myth of the American "frontier", pointed out that frontier settlers were not the avant-garde of urban settlement but represented the "expanding edge of the boosters' urban empire" (Cronon 2009, 51), then urban migrants can be seen as the obverse of the settler: occupants of an in-between space which expands the city back into the countryside, physically, economically and culturally.

Both as "arrival city" and as "factory of the future", the megalopolis, then, poses one of the key challenges for the social sciences and cultural theory in the twenty-first century.

So far, the transitory urban spaces of the global "arrival cities" have drawn three main responses from Western observers. From a classical Marxist point of view, they appear as harbingers of a "planet of slums" and further proof of the theory of impoverishment (Davis 2006). In sharp contrast to such perspectives, urbanists and architectural theorists like Rem Koolhaas[1] have celebrated transitory urban spaces in vitalist terms as self-organizing laboratories of coming modes of urbanity. And finally, to neo-liberal and libertarian observers, slums like Dharavi in Mumbai are exemplars of free-trade zones in which industry and trade flourish not in spite, but because of the complete absence of government regulation and top-down policy frameworks (The Economist 2007).

With his concept of "arrival cities", Doug Saunders has developed a more nuanced and less normative reading of these spaces and the legal frameworks and thresholds, economic dynamics and social conflicts which shape them. But as the work of scholars like Ravi Sundaram or anthropologist Brian Larkin shows, a media studies perspective is required to fully understand the new urban spaces, and the role of cities as the "factories of the future".

This turn to a media urbanity framework aligns with the current transformation of film and media studies. Scalable across a variety of formats and standardized in view of global circulation, the moving image has always been both an image of movement and an image on the move.

[1] See Bregtje van den Haak's documentary on Koolhaas' Lagos trip for Dutch television, released in 2003: https://wiki.beeldengeluid.nl/index.php/Lagos/Koolhaas.

However, for a long-time film studies defined its object in terms of the cinema hall and the classical dispositive of projection, despite the fact that small-gauge formats made moving images available for the home and other non-theatrical settings. But starting with VHS technology in the 1980s and particularly with VCD and DVD in the 1990s, moving images began to appear everywhere outside of the cinema. One response in film and media research was from media ontology to what we might call media onto-topology: a shift from the question "What is cinema?" to another question, "Where is cinema?". This led to heightened attention to home video formats and digital platforms, but also to the emergence of an entire new field of study on non-theatrical films, ranging from home movies to industrial, educational and scientific films. However, if the initial focus of research responding to the "Where is cinema?" question was largely on Western middle-class settings and technologies, and on the highly stratified global value chains of American and European culture industries with their attendant legal and policy frameworks, the study of media urbanity and the urban media economies and ecologies of megacities has shifted perspectives and created a whole new set of challenges, not least in terms of methodology.

To begin with, megacities like Lagos, Karachi, Delhi, Manila or Seoul challenge modern notions of order and visibility in urban design. Modern cities of the eighteenth through the twentieth century are usually organized around carefully planned axes of vision anchored in landmark buildings and iconic sites—think of the Paris of Haussmann, L'Enfant's plan for Washington D.C. or Lutyen's plan for New Delhi. Nineteenth-century urbanism was strongly influenced by miasma theory, a now obsolete medical theory which claimed that stagnant bad air in dense urban settings was the main factor in the spread of infectious disease. This assumption prompted urban planners to promote open spaces with well-regulated air currents and even build houses on stilts where possible, not least in the tropical settings of the parts of India and Africa colonized by European imperialists (Curtin 1985, 600). Contemporary megacities, on the other hand, consist in large part of unplanned and unregulated developments, the mere sight of which would cause miasma theorists nightmares, and which are either destined to be razed, absorbed into more regulated spatial layouts or continue to graft themselves precariously onto existing infrastructures. If what Frank Lloyd Wright described as centripetal sprawl in the 1950s works as the extension of urban space into the well-ordered suburban space of the nineteenth- and twentieth-century garden city, the sprawl of the megacity appears to be amorphous and chaotic, at least at first sight. It poses a problem of intelligibility, certainly for the external observer, and sometimes for those who inhabit this space.

To get a sense of the vast expansions of one of the world's new mega cities, Lagos, Rem Koolhaas famously took a helicopter ride in the early 2000s. Koolhaas made his name as a theorist with his embrace of the big city as a metabolism and the "chance-like" nature of city life in the late 1970s. In the early 2000s, as the convenor of the "Project on the City" at Harvard's Graduate School of Design, he turned his attention to new forms of urban development in China and Sub-Saharan Africa, with Lagos as one of his main points of interest. While Koolhaas did spend time on the ground in various areas of the city, his choice of the helicopter as the preferred mode of capture of the city's layout and structure is significant. The aerial view carries the promise of restoring at least a semblance of the visual order of the modern city. However, considering the complexity and unwieldiness of the city's layout, the best that the view from the mobile platform of the helicopter can offer is a view "at a glance", comparable to that of a commander on a battlefield, who has to make decisions with incomplete information and blinded by what Clausewitz called "the fog of war", or comparable to a manager looking at a chart, which is a poor rendering of the reality but serves as much to reconfirm the manager's decision-making authority as it helps to make that decision (Hoof 2020). It suits the epistemology of the decision-maker's glance that the viewing platform is mobile: moving over and across the city in a helicopter resonates with an understanding of the city as a complex ongoing, differential process rather than a static layout and a normative representation of social order and power relations. Contrast Koolhaas' helicopter ride with the way millions of Lagosians experience their city on a daily basis: through the windows of one of the yellow 1980s VW Transporter buses which serve as public transport operated by private owners and traverse the city on routes announced by the ticket seller/conductor on board at each stop. It is also a form of mechanical mobility, but one in which the lateral gaze through the car window replaces the vertical gaze from the helicopter. This lateral gaze offers a procedural view in which the duration of the trip rather than the horizon of what is visible at a glance from above is the measure of urban space. But the contrast in temporality concerns not just the difference between instantaneity and duration. The bus ride is usually a commute, a routine occurrence, which is one of the "minor" practices in the sense of De Certeau which constitute the fabric of urban life (De Certeau 2002; Sundaram 2011, 171). The architect's view is a view of the city as it is, but a view directed towards the future. It serves to anticipate what cities, and not just this city, will become, but it also re-defines the architect's and urbanist's role from planner to a more cybernetic understanding of observer-cum-facilitator, a participant in a process which moves towards outcomes which can better be gamed than planned. What is important to understand is that the urban space of the architect in the helicopter and the commuter on the bus is both the same and not the same. It is a boundary object, an infrastructure which is, as Susan Leigh Starr puts it, "both relational and ecological" and means different things to different people (Leigh Starr 1999). Rather than a homogeneous public sphere, this urban space is the site of starkly different temporalities and with it, of multiple "temporal publics" in the sense of Sarah Sharma. Based on her micro-analysis of the spatiotemporal arrangements of such work environment as airports, in which multiple states of temporal being coexist—ranging from passengers with time on their hands to service personnel working on tight schedules—Sharma offers a critique of what we may describe the spatial bias of established notions of the public sphere:

> While we may share space and time in a certain snapshot of social life or in the crossing of trajectories, on another fundamental level the dream of shared space occurs at the expense of acknowledging the uneven time politics that underlie coming together. More often than not, the sharing of space and a moment in time with one another is part of a synchronic relation of power. (Sharma 2014, 146)

What Sharma proposes instead is a "power-chronography", an analysis of the power relations in specific spatiotemporal settings. As the main tool of power-chronography the concept of the temporal public, a public defined in spatiotemporal rather than just spatial terms, offers a perspective on social politics as an intersection of the geopolitical (space) and the chronopolitical (time), a perspective which allows us to understand what exactly it is that brings together and separates the architect in the helicopter and the commuter on the bus, or the traveller in the airport and the staff making the airport run. What is more, combined with Leigh Starr's concept of infrastructure as boundary object which means different things to different people, Sharma's notion of the temporal public provides a handle not just on the complex dynamics of the transitional spaces of megacities, but also on narratives, props and infrastructure as configurations of media urbanity.

Props and/as Infrastructure

Props are material objects in and for narrative worlds. Their actual and purported materialities may coincide, but do not have to. They serve purposes ranging from "world-making detail", to quote Roland Barthes' term for descriptive excess in nineteenth-century literature (Barthes 2005), to McGuffin, Hitchcock's term for an object which is the focus of and motivates the actions of a film's characters and drives the narrative but can be entirely devoid of meaning, or material substance, for that matter. For the most part, props are simply part of the character's environment, present in the mode of what Heidegger calls "Zuhandenheit", the being-at-hand of material objects, which may or may not serve as tools or for other purposes. Some props carry symbolic meanings, like statues in films, which are decorations indicating traits of characters in the film or speak to film more generally as an art of gesture. Other props acquire specific symbolic meaning in the course of the narrative. In Hou Hsiao-hsien's "A Time to Live and a Time to Die" from 1985 a refugee and, in his mind, temporary exile from mainland China spends a good part of his life sitting in a bamboo chair and ends up dying in it. Upon his death, his children find a note which reveals that their father had bought the cheap bamboo furniture because he expected to leave it behind upon his imminent return to the mainland. The chair turns into the symbol of a life tragically spent in a permanent state of transition (Hediger 2020). For the most part, however, props do not overtly speak to their latent meanings. For instance, coffee is a staple in American Western films, part of the narrative world's supply system, so much so that drinking coffee comes like breathing to most Western heroes. A closer analysis reveals that coffee serves a specific purpose. Different from whisky and water, coffee brings people without previous ties together in what we may describe as a regenerative form of sociability, one in which past ills and failings are forgotten and forgiven for a new community to emerge. It is a secular sacrament of community, or more specifically of a community of white settlers, since Native Americans never share in the sociability coffee affords, nor do African Americans, who are largely absent from the films anyways (Hediger 2021). The meaning of the bamboo chair can be attributed to the mastery of Hou, but coffee in the West is a part of a narrative system which transcends individual works and functions independently of an author's intentions, personal style or the value accorded to the work by critics.

Understood in that sense we can see that props share some characteristics with infrastructure. They can be different things to different people, and an individual prop can serve different purposes according to the occasion. More specifically, props share with infrastructures the characteristics of reliance—they usually perform the tasks for which they were built; invisibility and breakdown—infrastructures are "as ordinary and unremarkable to us as trees, daylight and dirt" (Edwards, quoted in Lobato 2019, 76), that is much like most props, and they only stand out in the case of malfunction, which is one way in which props become drivers of the narrative; codetermination—infrastructures, like props in their being-at-hand, shape patterns of experience and define which actions are possible and which are not; and standardization—infrastructures rely on consensus about equipment and formats; props must be immediately recognizable in their functionality to serve their narrative purpose (Lobato 2019, 76–77).

Props, then, can be understood as elements of the infrastructure, so to speak, of narrative worlds. That includes media and media as infrastructure. Film has parallel histories with railways, the telegraph and the telephone (Martin 1991; Kirby 1997), and entire book could be written about telephones, printing presses and surveillance technologies in the films of Fritz Lang alone. By using infrastructure as props, narrative film has long worked to make the material and technological arrangements of modernity intelligible. Non-fiction films have done so even more explicitly; the process film, which typically details the industrial production of material objects for mass consumption, is one of the most consistent and popular non-fiction film genres (Skvirsky 2020). What is of interest here is the role of infrastructures as props which make the complex spatiotemporal arrangements of twenty-first-century megacities legible. If the city does indeed replace the factory as the most important site of value creation in the twenty-first century, as Richard Baldwin argues, the narratives on and of media infrastructures are the process films of contemporary media urbanity. At the same time, if we take a clue from the role coffee plays in American Westerns, props, including media infrastructures as props, provide a key to social politics and forms of community and to the power-chronography of temporal publics as they relate to these infrastructures.

What follows now is a brief discussion of the three examples which offer a possible map for further inquiry. It is a tour d'horizon which bears some resemblance with Koolhaas' helicopter ride: A synopsis at a glance, rather than an in-depth inquiry on the ground. Hopefully, it will not preclude, but lay the groundwork for such an inquiry.

Multiple Screens and Questions of Unity: "Up North" (Tope Oshin, Nigeria 2019)

In a recent study economic historian Werner Plumpe argues that modern capitalism and the industrial revolution can be explained through the emergence of large markets for mass consumption in combination with supply-elastic labour and the requisite technology in Northwestern Europe in the seventeenth and eighteenth centuries (Plumpe 2019; incidentally, Plumpe does not mention the plunder of Bengal, which arguably provided the capital for the industrial revolution). Mass consumption, in other words, is not a toxic side effect of capitalism and industrial development, but its driver. Proof of Plumpe's argument, together with Baldwin's claim that cities replace the traditional brick-and-mortar factory as the factory of the future, can be found in the Nigerian film industries since the early 1990s. Nigeria had a domestic film and television production funded through the National Television Authority and, starting in the early 1980s, the Nigerian Film Corporation, which came to a standstill due to austerity measures imposed on the country by the IMF in the late 1980s (Ekwuazi 2018). In the early 1990s, Kenneth Nnebue, a trader for electronic goods in Lagos stepped into the void and started producing video films with local content to drive sales for imported Taiwanese VHS recorders. Living in Bondage, a two-part, six-hour video film was marketed on VHS tapes carefully packaged in transparent plastic foil go give it a more legitimate industry look and turned into a major hit (Haynes 2016). It paved the way for the emergence of a film industry focused on what Alessandro Jedlowski proposes to call "small screen cinema" (Jedlowski 2012): Video films produced primarily for domestic consumption modelled on stylistic templates ranging from Brazilian telenovelas to Bollywood films (Krings 2015) and tackling stories of local and regional interest and produced in English or one of the other national languages of Nigeria, primarily Igbo, Yoruba and Hausa (McCain 2013). Centring on Lagos for the English, Yoruba and Igbo films and on Kano, the main urban centre of Northern Nigeria, for Hausa language films, what is now colloquially known as "Nollywood" quickly became one of Nigeria's main exports. Satellite channels like the South Africa-based "Africa Magic" screen Nollywood films all across sub-Saharan Africa around the clock (Ekwuazi 2014). Irkokotv.com, a streaming platform and low-price Subscription Video on Demand Service (SVOD), reaches Nigerian and West African diasporas across the world. Increasingly Nollywood films appear on Netflix, with noted directors like Kunle Afolayan signing multi-picture deals directly with the American streaming service, which keeps its African headquarters based out of Amsterdam in the Netherlands (Agina 2021). The layered ecology of streaming thus increasingly complements and supersedes earlier distribution practices, which relied on what Ramon Lobato has called "Shadow Economies of Cinema" (Lobato 2012). A prime example of what Vasudevan calls "pirate modernity", Nollywood had first relied on informal networks for the distribution of VHS, VCD and DVD copies of film which provided Nigerian audiences with access to the global market for cultural goods in pirate form. These networks were repurposed to further the growth of the local film industry and created revenue streams for producers even in the absence of regimes of copyright enforcement (Larkin 2004; Lobato 2010). Nollywood films appearing on Netflix, then, mark the industries' progressive transition to more formalized, contract-based production and distribution practices, and from the informal physical infrastructures of urban space to a hybrid of urban space and digital networks.

Tope Oshin's romantic comedy "Up North" exemplifies this transition both in its production and distribution history and in its narrative. A hit upon its original cinematic release through Nigeria's still limited network of multiplex cinemas in urban centres, "Up North" is one of a package of Nollywood films produced by independent filmmakers in 2014 and acquired by the streaming service as part of a package intended to test the reach of Nollywood films with the services relatively well-to-do subscriber base (in Nigeria, a Netflix subscription is more than ten times more expensive than an Irokotv.com subscription, and like in India, Netflix Naija mostly addresses an educated English language middle-class audience, including Nigeria's sizeable population of university students). Along with director, producer and actress Genevieve Nnaji or producer and television magnate Mo Abudu, Tope Oshin is one of the women filmmakers currently redefining an industry which had long been catering to a primarily female audience. "Up North" is the story of Bassey Otuekong (Banky Wellington), the scion of a rich family of Lagosian entrepreneurs who is forced to trade his jet-setting life for his mandatory year of national service in the Nigerian federal government's National Youth Service Corps programme for university graduates. A fixture of Lagos nightlife, the young MIT engineering graduate is sent to the Northern rural state of Bauchi to work as a fitness instructor in a primary school. An encounter with the beautiful, but stern schoolteacher Maryam (Rahama Sadau) changes his life. He falls in love, leaves his worldly lover in Lagos and ultimately chooses to trade city life with a career as a schoolteacher in the rural North. "Up North" shares a key narrative concern with Genevieve Nnaji's 2018 film "Lionheart", which premiered at the Toronto Film Festival and is also on Netflix. "Lionheart" and which tells the story of an Igbo bus entrepreneur (Nnaji) from the South East who fights for the survival of her father's company and is saved only through a merger with a Northern competitor, which also results in her marriage with her new business partner's son. Both "Up North" and "Lionheart" tell love stories as stories of national unity. Like most African countries an outcome of the power sharing agreement negotiated between European colonial powers at the Berlin conference of 1883. Created by fiat of the British from two protectorates encompassing previously independent polities in 1914, including the Northern Sokoto caliphate and the Benin kingdom in the South, the geographical entity now known as Nigeria is one of the most linguistically and culturally diverse countries in the world, with the Muslim Hausa and Fulani communities in the North and the (largely) Christian Yoruba and Igbo communities in the South as the three dominant ethnicities. Described at one point by Northern political leader Ahmadu Bello as "the mistake of 1914", Nigeria has struggled to maintain its unity through the Biafra war, in which the Igbo tried to form an independent nation state after the military coup of 1966 (Siolun 2009; Offodile 2016), and is currently under strain from the Boko Haram insurgency, which runs across the country's North East. The country's unity has long been predicated in good part on the integrative force of its military, one of the few institutions in which sectional identities are muted by functional hierarchies, which helps explain that even after the now seemingly permanent transition back to democracy in 1999, the country has largely been governed by former generals and military dictators turned proponents of democracy (Olusegun Obasanjo from 1999 to 2007 and Muhammadu Buhari since 2015). Against this backdrop films like "Lionheart" and "Up North" acquire an additional layer of meaning as they tell stories of love which bridge and conquer all the polar opposition which constitute and constantly threaten to tear apart Nigeria as a nation state: North versus South, Muslim versus Christian, Rural versus Urban, Hausa/Fulani versus Yoruba and Igbo. As one critic for a Lagos publication writes: "Oshin's film is awake to the lack of opportunities open to northern women; the place of religion and the Sharia law, the favouring of the male child over the female" before adding: "There is even a passing nod to the insurgency in the north. See whether you can spot it" (Kan 2019). But ultimately love conquers all. In

"Up North", like in "Lionheart", falling in love is an act of community formation.

"Up North" tells this story as a layered narrative, combining conventional mise-en-scène with a double screen aesthetic in which a steady stream of text messages and Instagram posts pop up over the image. The messages and posts detail the exchanges between the protagonist, his girlfriend, his family and later his new love in the North. Sometimes they serve as asides and running comment, but they often convey important insights on plot and character. The messages and posts set the pace of the narrative and visualize the character's dilemma of being in several places at once, or being torn between places. But if digital communication initially keeps the protagonist dispersed, it ultimately also serves as the catalyst of the unity formed through the bond of love. The messages and the posts on the screen gesture towards digital communication as the infrastructure through which the diverging temporalities of rural life in the North and the fast-paced lifestyle of the Southern business community can come together. But the happy end cannot gloss over the fact they ultimately remain apart. Nor is it trivial that the digital platforms the protagonist uses are, so to speak, borrowed infrastructures: branded Silicon Valley tech platforms which supplement local and regional infrastructures as they do now in Nigeria as in many, if not most, parts of the world. If the question is one of social politics and community formation, one thing that "Up North" helps to understand is that the temporality of borrowed infrastructures is also one of borrowed time.

No Politics Without Media: Lucifer (Prithviraj Sukumaran, India 2019)

When Jean-Luc Godard received the lifetime achievement award of the Kerala Film Festival in early 2021, he agreed to a wide-ranging, ninety-minute interview via Zoom with Indian film scholar C.S. Venkiteswaran. Speaking from his home in Rolle, a small Swiss town on Lake Geneva, next to his native Nyon, where Godard has lived and worked with his wife and partner Anne-Marie Miéville since the 1970s, the ninety-year old director opened the proceedings with a question: "Where, geographically, is Kerala?" Rather than being offended, the moderator instructed Godard about Kerala's location on India's southwestern tip and pointed out that India is "one of the most well-developed states in the Indian Union", adding that "film culture is very much a part of this, literacy and education". Godard had traded Paris for his native pastures in the early 1970s, when Maoism was all the rage amongst French intellectuals and immersion in rural communities was the order of the day (although the Swiss shore of Lake Geneva is as far from rural poverty as you can get), and he named his production outfit "Peripheria Films", to mark his distance from Paris' eight arrondissement where all French production firms are headquartered. Kerala, together with Bengal the only place in the world where communist parties regularly win, and lose, free elections and accept the results, could have caught Godard's attention, and he could have looked it up on the internet. Instead he chose to be honest and show his ignorance, demonstrating the persistence of twentieth-century mental geographies of cultural production in the process. Periphery, for Godard, still means to have his eyes firmly trained on Paris.

Kerala, a state the size of Belgium but with four times its population at 33 million inhabitants, sometimes feels like an uninterrupted conurbation to the visitor driving through it by car, with reminiscent of Germany's Ruhr valley or Southern California in its density. It is home to a flourishing publishing industry and the world's largest newspaper by size of penetration relative to the potential readership, the daily *Malayalam Manorama* with a daily readership of ten million. It is a linguistically homogenous but religiously diverse state, with Muslims making up for 25% and Christians for 18% of the population, India's largest share of Christians in any state (mostly Roman Catholics, but also 2.4 million Syrian Catholics, a denomination which can be traced back to the fourth century AD on the Malabar coast and which include the Gospel according to Thomas in Scripture). It is also home to India's fourth largest regional film industry, after the Hindi film industry popularly known as Bollywood and the industries of Bengal and Tamil Nadu. A source of considerable regional pride and firmly anchored in the state's audiences graces, "Mollywood" is known for realistic storylines and advanced visual styles in films which often reference art cinema as much as they align with more formulaic templates of Hindi mainstream films. Shaji N. Karun's 1999 film Vanaprastham ("The Last Dance"), which was coproduced by French writer Pierre Assouline and the film's star—and Malayalam cinema superstar—Mohanlal, remains to this day one of the few films from a regional Indian film industry outside of Bengal to win a major award at an A-list festival, for whatever such validation by Western gate keepers of film culture is worth.

Contemporary Malayalam films are available through streaming services like Amazon Prime and, of course, as part of the inflight entertainment programme of the various Gulf State airlines, which, as a reflection of their route networks and clientele, have some the most comprehensive Asian—and African, for that matter—cinema screening lists anywhere (the Gulf States invested heavily in airlines with a view to the post-fossil fuel economy because two thirds of the world population live within an eight hour flight radius of the Gulf States). One important source for Malayalam films, but also other regional Indian films, is Einthusan, a streaming profile which boasts a catalogue of more than four thousand films and currently offers subscriptions for 99 (!) years for the equivalent of about 25 Euros.

"Mollywood" is located in Kochi, Kerala's largest city and the burial place of Vasco da Gama, whose "discovery" of the naval route to India in 1499 set the stage for Europe's naval-based imperialism and the later colonization of India. Kochi is the site of one of India's largest public transport projects, the Kochi metro, which started construction in 2013 and had its first line opened to the public in 2017. Kochi is also the site of what is now one of the most visited public venue in India, more popular even than the Taj Mahal, the LuLu International Mall, which is directly connected to the metro and located near the intersection of its two main lines. LuLu Mall also includes a multiplex cinema showing Malayalam, Tamil, Hindi, Bengali and Hollywood films alongside each other.

Despite being among the "most developed states in the Indian Union", Kerala is also a major site of migration. Apart from its share in the Indian diaspora across the world and particularly in the United Kingdom, Kerala provides 2.2 million migrant labourers to the workforce in the Gulf States, the largest contingent of from any state in the Indian Union (Menon 2020). Gulf States migration is an important topic in Malayalam literature and cinema, both in terms of depictions of the hardships of migratory life and in the sense that the Gulf presented a template of a desirable modernity which was reflected in the aesthetics of Malayalam films since the 1970s (Radakrishan 2009). Bengali novelist Amitav Ghosh coined the term "petro-fiction" for literary works portraying migrant labour in the Gulf oil industries, a term which Priya Menon has recently been expanded to include all forms of fiction dealing with Gulf state migration in Kerala, for example the work of writers like Benyamin, Deepak Unnikrishnan and others (Menon 2020).

The action-thriller "Lucifer" from 2019, the directorial debut of action star Prithviraj Sukumaran and currently the highest grossing Malayalam film of all time, qualifies at least partially as a work of petro-fiction in that broader sense. The title character Lucifer, played by Mohanlal, is a mysterious character who previously worked in post-war reconstruction in Iraq and Afghanistan and was involved in smuggling gold from the Gulf to Kerala. His birth name is Stephen Nedumpally, a name which marks him as a Christian, and in the film he is at the heart of a complicated political intrigue: The chief minister of the state, P.K. Ramdas or PKR, dies while in treatment in a hospital owned by the daughter of the leader of the opposition. Sensing an opportunity for their party in the upcoming election, PKR's handlers stage a riot of the TV cameras in front of the hospital, even as the late chief minister's children, daughter Priya and son Jathin, line up for their late father's succession. Jathin would prefer to stay in the United States with his family, and Priya temporarily gains the upper hand with the help of her ambitious second husband Bobby (played by Bollywood star Vivek Oberoi), who she incidentally encountered after her first husband's death by car accident in Dubai seven years ago. But then Stephen/Lucifer returns from the Gulf and enters the scene. Little is known about his background other than he was close to PKR and put in line for his succession at some point six years ago. The mere arrival of Stephen prompts the formation of a spontaneous political movement outside of party structures, a textbook illustration of Partha Chatterjee's "politics of the governed", and a fierce battle for power ensues. As it turns out, Bobby is part of an international gang of drug dealers and plots to take over the government to make the state the home base of his crime syndicate. Stephen, who runs an orphanage on the side and is revealed to be PKR's illegitimate son, ultimately prevails but relinquishes the reins of power in favour of Jathin, who reaffirms his party's control over the state government. Stephen returns to Afghanistan, in search of new challenges (and in preparation of the next instalment of what is explicitly designed as an action film franchise).

Apart from the pervasive Christological visual metaphors surrounding the title character, which are unique to Malayalam films in Indian cinema, what makes the film distinctive is the framing of the story as a thoroughly mediate narrative. From the outset the narration foregrounds technologies of observation and transmission, starting with the filing system of Interpol and the CIA, in which Lucifer/Stephen under his alias Kureshi Abra'am, and moving on the television cameras for which PKR's handlers stage a riot, which they direct over mobile phones in front of the hospital, and on to the Facebook live stream of Govardhan, an independent researcher who publishes a dossier on a secret cabal to control the Indian government, which sounds like a typical Facebook conspiracy theory but turns out to be largely true as the film progresses, and further on a team of television journalists who are partly complicit in Bobby's plot. The fight for political power is a fight over mediated perceptions, in which backstage and front stage are never clearly distinguishable and all roles and identities are ultimately elaborate performances. Narratives of corruption are a staple of Malayalam cinema, and Mohanlal's role as the upstanding, uncompromising fighter for the integrity of government and its institutions is one he reprises time and again. But Mohanlal's filmography also includes a memorable turn as a film star-turned-chief minister in Mani Ratnam's Tamil film "Iruvar" from 1997, a thinly veiled biopic of former Tamil Nadu chief minister M.G. Ramachandran. Much like in "Iruvar", in Lucifer he is himself a political player, but a player in a political mirror cabinet who ultimately remains a mystery even to those closest to him. If Partha Chatterjee appears to think of social politics as "politics before media", as Ravi Sundaram points out, a film like Lucifer assures that there is no politics, whether party politics, or the social politics of the governed, before and outside media. At the same time the use of media infrastructures as props in Sukumaran's film reveals a complex layering of temporal publics and shows that a demos fractured by class, caste and religious identity, and dispersed in the spaces of migration, cannot be easily recomposed in a coherent mediated public sphere, despite the best promises of connectivity of digital communication networks. Which is also why the franchise format, which the film's ending implies and which the director's public statements promise, is appropriate: the political action thriller captures a field of political action and popular politics in which all gains are at best temporary. As Vasudevan writes of the pirate media city, it "mixed debris, recycled structures and hyper-modern technologies in its appropriation of media infrastructures, refusing the progressive determination of its actions". It is possible to read a film like Lucifer as a performative enactment of just such a challenge to the ostensible rationality of liberal democratic governance.

Welcome to the Digital Ghost Town: "At Dusk" (Hwang Sok-yong, South Korea 2015, Trans. 2019)

South Korea can be described as a model case for most political, economic, social and cultural issue of the last fifty to seventy years. A country with a four-thousand-year history, including 45 years of colonial history, a compressed modernity with rapid industrialization intertwined with a convoluted, often violent democratization process, and now a global player in technology and culture (Cumings 2005). "Hallyu", the Korean wave in pop music, television drama and film, was first a regional phenomenon in East Asia. The unexpected success of the 2002 television series Winter Sonata in Japan was an early indicator of things to come. "Gangnam Style", a 2012 social satire song by rapper Psy making fun of nouveau rich denizens of Seoul which became the first YouTube video to register more than one billion views, was a breakthrough moment. Now, K-Pop bands like BTS command a global following of loyal fans through YouTube[2] (Lee and Nornes 2015), and Korean films like Bong Joon-ho's Parasite (2019), which had become regular entries in major film festivals since the early 2000s, now win major awards in Europe and the United States, even as they continue to easily compete with Hollywood

[2] https://youtu.be/nwquFZlitTU.

blockbusters in the domestic market. While K-Pop and television dramas tend to offer sanitized entertainment which is designed to offend no one in the interest of exportability, reflections on the more violent and contentious aspects of recent Korean history appear to be the preserve of literature and film. Particularly since the early 2000s, directors like Park Chan-wook, Lee Chang-dong, Kim Jee-woon, Im Sang-soo and Kim Kyung-chan have progressively addressed key aspects of Korea's modern history in films ranging from the colonial occupation by the Japanese from 1900 to 1945 to the Korean War, the Gwangju massacre committed by the military dictatorship in 1980 and the 1987 student process which were a turning point in the democratization process (Kim 2011; Choe 2016). In South Korean literature Hwang Sok-yong, who was born in 1943, has been a consistent voice in articulating various aspects of Korean history since the early 1970s. A labour activist and political prisoner in the 1960s, he served as part of the Republic of Korea Marine Corps alongside American soldiers in the Vietnam War from 1966 to 1969, an experience which he recounts in his 2008 multi-character novel Evening Star. Hwang's literary breakthrough came in 1970 with Mr Han's Chronicle, the story of a family separated by the Korean War. Hwang was jailed again in the 1990s following a very public visit to North Korea, which at the time and before President Kim Dae-jung's policy of détente was legally considered an act of treason. Hwang's biggest success came with The Old Garden in 2000, the story of a democracy activist and a painter whose paths cross briefly only but who remain to each other for the rest of their lives, a novel which draws on Hwang's second experience in jail and was turned into a film by Im Sang-soo in 2006. In a recent study of Im's film and Park Kwan-su's 1995 film A Single Spark, about the self-immolation of garment worker Jeon Tae-il, which provoked massive labour unrest in the early 1970s, Jecheol Park shows how multilayered temporalities are a signature trait of contemporary Korean films (Park 2019), reflecting the conditions of compressed modernity. In his 2015 novel *At Dusk*, which was published in English in 2019, Hwang Sok-yong connects the problem of multilayered temporalities to question of both physical and digital infrastructure. The protagonist of the novel, Park Min-woo, is a successful architect who rose from poverty in a miserable periphery of Seoul to a position of power and wealth. He rode the wave of development which has taken South Korea from agrarian colony to postindustrial wealth in the short space of seven decades, but as an architect who made a good share of his fortune transforming the transitional spaces of the "arrival city" into lucrative developments, he is also a protagonist of this transformation. When his company becomes embroiled in a corruption scandal, he begins to reflect on his role in the city and the country's rapid transformation. At the same time, he starts receiving messages from Chan Soo-na, a woman he grew up with and once loved but later betrayed and left behind. Her text and voice messages, accompanied by copies of letters and diary entries, force him to confront his past and the way in which he detached himself from his humble origins by marrying into a wealthy, well-connected family. As the story progresses, another protagonist emerges: A young woman who trained as an actor and theatre director but now works as a store clerk, trying to recover from the failure of her first own production. We soon realize that the young performer had befriended Chan Soo-na. As the pieces of the story fall into place in their chronological order we further understand that Chan Soo-na has actually died before the story starts, and that the young woman has used voice recordings and her friend's diaries and letters to stage a performance in which Park Min-woo interacts not with his old lover, but her spectral double: with a digital ghost. Park's remembrance turns out to have been a digital séance of sorts, and the success of the young woman's performance incidentally leads to his downfall. In Hwang Sok-yong's novel, the ontology of media gives way not just to an onto-topology, but a hauntology of media. One of the promises of modernist urban planning and development particularly in post-colonial cities like Seoul had been the elimination and the emancipation from the past and its rural traditions. If endless proliferation as the primary mode of urban development was "unleased with globalization rather than from non-modern pasts, as had been feared during the 1950s" (Vasudevan 2011, 174), then Hwang's narrative dramatizes the persistence of those pasts in and through digital infrastructures. And so, at least for a fleeting moment, the twenty-first-century megalopolis turns out to be a digital ghost town.

References

Agina, Añulika. 2021. Netflix and the Transnationalization of Nollywood. In *Post45. New Filmic Geographies*, ed. Susanne Ezerink. https://post45.org/2021/04/netflix-and-the-transnationalization-of-nollywood/.

Baldwin, Richard. 2016. *The Great Convergence. Information Technology and the New Globalization*. Cambridge: Harvard University Press.

Barthes, Roland. 2005. The Reality Effect. In *The Novel: An Anthology of Criticism and Theory, 1900–2000*, ed. Dorothy J. Hale, 229–234. Hoboken: Wiley.

Bhutto, Fatima. 2019. *New Kings of the World. Dispatches from Bollywood, Dizi and K-Pop*. New York: Columbia University Press.

Chatterjee, Partha. 2004. *The Politics of the Governed. Reflections on Popular Politics in Most of the World*. New York: Columbia University Press.

Choe, Steve. 2016. *Sovereign Violence. Ethics and South Korean Cinema in the New Millennium*. Amsterdam: Amsterdam University Press.

Cronon, William. 2009. *Nature's Metropolis. Chicago and the Great West*. New York: W.W. Norton.

Cumings, Bruce. 2005. *Korea's Place in the Sun. A Modern History*. New York: W.W. Norton.

Curtin, Philip D. 1985. Medical Knowledge and Urban Planning in Tropical Africa. *The American Historical Review* 90 (3): 594–613.

Davis, Mike. 2006. *Planet of Slums*. London: Verso.

De Certeau, Michel. 2002. *The Practice of Everyday Life*. Trans. Timothy J. Tomasik. Berkeley: University of California Press.

Ekwuazi, Hyginus. 2014. The Perception/Reception of DSTV/Multichoice's Africa Magic Channels by Selected African Audiences. *Journal of African Cinemas* 6 (1): 21–48.

———. 2018. *Film in Nigeria*. Trenton, NJ: Africa World Press.

Engel, Ulf, Marc Boeckler, and Dieter Müller-Mahn, eds. 2018. *Spatial Practices: Territory, Border, and Infrastructure in Africa*. Leiden; Boston: Brill.

Fuhrmann, A. 2016. This Area Is [Not] Under Quarantine. Rethinking Southeast/Asia Through Studies of the Cinema. In *Area Studies at the Crossroads. Knowledge Production After the Mobility Turn*, ed. K. Mielke and A.-K. Hornidge, 251–268. London: Palgrave Macmillan.

Haynes, Jonathan. 2016. *Nollywood. The Creation of Nigerian Film Genres*. Chicago: Chicago University Press.

Hediger, Vinzenz. 2010. Infectious Images: Haneke, Cameron, Egoyan and Dueling Epistemologies of Video and Film. In *A Companion to Michael Haneke*, ed. Roy Grundman, 91–112. Hoboken: Wiley-Blackwell.

———. 2020. The Ephemeral Cathedral. Bodies of Stone and Configurations of Film. In *Bodies of Stone in the Media, Visual Culture and the Arts*, ed. Alesseandra Violi, Barbara Grespi, Andrea Pinotti, and Pietro Conte, 106–125. Amsterdam: Amsterdam University Press.

———. 2021. Schwarzer Kaffee für weiße Siedler. Der amerikanische Western als Nachkriegsfilm. In *Politik der Grenze. Interdisziplinäre Perspektiven auf die Frontier im Western der Gegenwart*, ed. Anja Peltzer and Jörn Ahrens. Cologne: Herbert von Halem.

Heise, Ursula K. 2017. Planet, Species, Justice – And the Stories We Tell About Them. In *The Routledge Companion to Environmental Humanities*, ed. Ursula K. Heise, Jon Christensen, and Michelle Niemann. London; New York: Routledge.

Hoof, Florian. 2020. *Angels of Efficiency. A Media History of Management Consulting*. Oxford; New York: Oxford University Press.

Jedlowski, Alessandro. 2012. Small Screen Cinema: Informality and Remediation in Nollywood. *Television & New Media* 13 (5): 431–446.

Kan, Toni. 2019, January 19. Up North: Tope Oshin's Thrilling Joy Ride. *The Cable*. https://lifestyle.thecable.ng/up-north-tope-oshins-thrilling-joy-ride/.

Kim, Kyung Hyun. 2011. *Virtual Hallyu. Korean Cinema in the Global Era*. Durham: Duke University Press.

Kirby, Lynne. 1997. *Parallel Tracks: The Railroad and the Silent Cinema*. Durham: Duke University Press.

Krings, Matthias. 2015. *African Appropriations: Cultural Difference, Mimesis, and Media*. Bloomington: Indiana University Press.

Larkin, Brian. 2004. Degraded Images, Distorted Sounds: Nigerian Video and the Infrastructure of Piracy. *Public Culture* 16 (2): 289–314.

———. 2008. *Signal and Noise. Media, Infrastructure, and Urban Culture in Nigeria*. Durham: Duke University Press.

Lee, Sangjoon, and Abe Mark Nornes, eds. 2015. *Hallyu 2.0. The Korean Wave in the Age of Social Media*. Ann Arbor: Michigan University Press.

Leigh Starr, Susan. 1999. The Ethnography of Infrastructure. *American Behavioral Scientist* 43 (3): 377–391.

Lobato, Ramon. 2010. Creative industries and informal economies: Lessons from Nollywood. *International Journal of Cultural Studies* 13 (4): 337–354.

———. 2012. *Shadow Economies of Cinema. Mapping Informal Film Distribution*. London: BFI.
———. 2019. *Netflix Nations. The Geography of Digital Distribution*. New York: New York University Press.
Martin, Adrian. 1991. On the Telephone. *Metro Magazine: Media & Education Magazine* 87 (Spring): 21.
Mattern, Sharon. 2017. *Code + Clay ... Data + Dirt. Five Thousand Years of Urban Media*. Minneapolis: University of Minnesota Press.
McCain, Carmen. 2013. Nollywood and Its Others. Questioning English Language Hegemony in Nollywood Studies. *The Global South* 7 (1, Spring): 30–54.
McLuhan, Marshall. 1964. *Understanding Media. The Extensions of Man*. New York: McGraw-Hill.
Menon, Priya. 2020, July. Kerala's Own Petrofiction: Literary Interventions in Gulf Migration Studies. Ala 23. http://ala.keralascholars.org/issues/23/keralas-own-petrofiction/.
Mukherjee, Rahul. 2019. *Radiant Infrastructures. Media, Environment, and Cultures of Uncertainty*. Durham: Duke University Press.
Offodile, Chudi. 2016. *The Politics of Biafra and the Future of Nigeria*. Morrisville: Lulu Publishing.
Park, Jecheol. 2019. Multitemporality between Sovereignty and Neoliberal Governmentality: A Single Spark and The Old Garden. In *Beyond Imperial Aesthetics. Theories of Art and Politics in East Asia*, ed. Steve Choe and Manyumo Inuoe, 243–262. Hong Kong: Hong Kong University Press.
Parks, Lisa, and Nicole Starosielski. 2015. *Signal Traffic. Critical Studies of Media Infrastructures*. Urbana: University of Illinois Press.
Plumpe, Werner. 2019. *Das kalte Herz. Kapitalismus: Geschichte einer andauernden Revolution*. Berlin: Rowohlt.
Radakrishan, Ratheesh. 2009. The Gulf in the Imagination: Migration, Malayalam Cinema, and Regional Identity. *Contributions to Indian Sociology* 43 (2): 217–245.
Ricoeur, Paul. 1990. *Time and Narrative. Vol. 1*. Chicago: Chicago University Press.
Saunders, Doug. 2011. *Arrival City. How the Greatest Migration in History is Reshaping Our World*. London; New York: Viking.
Schank, Roger C. 1995. *Tell Me Story. Narrative and Intelligence*. Evanston: Northwestern University Press.

Sharma, Sarah. 2014. *In the Meantime. Temporality and Cultural Politics.* Durham: Duke University Press.

Siolun, Max. 2009. *Oil, Politics and Violence. Nigeria's Military Coup Culture (1966–1976).* New York: Algora.

Skvirsky, Salomé. 2020. *The Process Genre. Cinema and the Aesthetic of Labor.* Durham: Duke University Press.

Sundaram, Ravi. 2011. *Pirate Modernity. Delhi's Media Urbanism.* London; New York: Routledge.

The Economist. 2007. A Flourishing Slum. https://www.economist.com/christmas-specials/2007/12/19/a-flourishing-slum.

United Nations World Urbanization Prospects. 2018. https://population.un.org/wup/Publications/Files/WUP2018-Report.pdf.

10

The In/Visible City: Cinema, Control and Contemporary Hong Kong

Rick Dolphijn

Introduction

Starting from the mid-nineteenth century, fiction authors began to write novels in which 'the other side of the city' was key. Unlike the sci-fi adventures that we had already seen with writers like Jules Verne and H.G. Wells, there were now quite some attempts (more philosophical, more political) to somehow capture how the city was in change on a more 'unconscious' level; what happened at the dark side of the city, the unseen part of the city, the city unheard of and so on. Two books need to be looked again. First, Fyodor Dostoyevsky, in his *Notes from Underground* ([1864] 2018), asks us to take a second look at life in St Petersburg, claiming that beneath its glittery surface, it was actually the most artificial and intentional city of his world. Later, Yevgeny Zamyatin's *We* ([1920] 1972), published in English translation in 1924, introduced us to the glass city of One State, a city set up and very much under control of a

R. Dolphijn (✉)
Utrecht University, Utrecht, Netherlands
e-mail: r.dolphijn@uu.nl

S. S. Das, A. R. Pratihar (eds.), *Technology, Urban Space and the Networked Community*, https://doi.org/10.1007/978-3-030-88809-1_10

new 'type' of government, a highly influential theme it turned out, notably to Aldous Huxley, George Orwell and the writers of that generation (Fig. 10.1).

Of course, both books (Dostoyevsky and Zamyatin) reflect the crisis of the city as these writers experienced it in their times. Dostoyevsky had just visited Paris in 1863 (a year before publishing his *Notes*), its centre being almost completely rebuilt by then (in a neo-classicist style), and the existentialist experiences of a city which now seemed to have been

Fig. 10.1 Sticker featuring Joker, decorating the streets of Hong Kong, 20 October 2019, photo by the author

covered up by uniform facades drive the narrative of the book. *Notes from Underground* makes us feel what it is like to live in such a 'modernized' city, a city which one does not recognize anymore, and in which one is also not recognized anymore. Its grand boulevards, its open construction, but also the people that walk its streets, show us how the 'facelessness' of the new city, its utopianism and denial of pain and fear, does not match with the lives of underground man and the people he engages with. But *Notes from Underground* is thereby not just a reflection upon the impossibility of life in the modern city in his day and age. This book talks just as well of the cities yet to come. This is the urban life not just of people of St Petersburg or of Paris; what Dostoyevsky foresaw has been realized in many ways; it is the new invisibility that we, city dwellers, all inhabit. It is the new invisible city we live.

Zamyatin gave us a reflection on the city of about sixty years later, a 'nearly-the-end-of-the-world' futurism, with an almost transparent city, a city of harmony and wonderfully organized, thanks to the strong presence of the state. Of course, the narrative unfolded here resonates with the way especially Soviet communist rule was able to redefine urban life ruthlessly, and how it managed to realize a new society in peace/fear. Zamyatin confronts us with a city of numbers; the schools, the butchers, everything was numbered (which indeed happened in Soviet times, and is still quite common in contemporary Russia). In the book even the main characters were numbered. Zamyatin presented us the datafied city (often strangely referred to in our days, as a 'smart' city). And the society he foresees is equally real to the one presented to us by Dostoyevsky. This time, however, the glass city is new visibility we, city dwellers, all inhabit. It is the new visible city we live.

The times of Zamyatin, the 1920s, were, even more so than the mid-nineteenth century, revolutionary. The many changes that redefined the city also gave it a new form of expression. A 'medium' which, compared to the written word, seemed much more integrated into city life itself, materially, ideologically, economically. Or perhaps I should say that this new form of expression matched the new and exciting cities that had rapidly reshaped urban life since the interbellum much better, on a global scale. Of course, I am referring to cinema now. The medium of appearance and disappearance. It is no coincidence that in the 1920s the work of Georges Mélies, an illusionist by training, was appreciated again, and considered to have given the moving image its foundation. Cinema had always been the medium of appearance and disappearance, of making things visible and of making things invisible.

During the interbellum, cities like Shanghai, Moscow, Berlin, Paris, New York and Chicago all exploded in terms of energy, size and height (with New York as its icon), giving the city a new fabric, a new image, a new feel. In no time, the idea of the city had little in common with the cities as we knew them from before World War I. And with this new city, we saw this new medium, cinema, gaining importance. Of course, there was a very good reason, that the prominent intellectuals of the days, from Bergson to Eisenstein, felt the urge to analyse this new medium for the masses, its technical and aesthetic possibilities and its political dangers: they understood that somehow, cinema mattered to how the city and its communities were in change, to how we witnessed the rise of a new city, a new life. These were proper philosophers, as they understood that "Philosophy is an anticipation of future thoughts and practices" (Serres and Latour [1990] 1995, 86). Cinema had to be analysed thoroughly, in the name of the future.

Published in German in 1925, Thea von Harbou's novel *Metropolis*, in many ways, combined the speculations of Dostoyevsky and Zamyatin. Situated in a uniform, technologically advanced city in the future, run by a very powerful government, von Harbou's protagonist who lives a decent life in the high-rise city falls in love with a person from the underworld, the dark and unknown parts of town, home to the lower classes. With flares of existentialist philosophy, and a strong spiritual, reflective tone, the book gives us a good idea what the novel can do, and how the city and the novel together, since the mid-nineteenth century, revealed the realities of social change, of class struggle and of the 'existential crises' that the whole process of modernization was unfolding.

Thea von Harbou was married to Fritz Lang, and together, two years after the publication of the novel, they finalized the movie *Metropolis* (1927). The movie was terribly expensive and didn't do very well in the theatres at first. But in the end, this movie would, in every way, reveal the new visibility/invisibility of the city differently.

Many Metropoles

In two ways, the movie *Metropolis* expressed the new ideas on the visible and the invisible city very differently from how this was done in the novel *Metropolis*, or even, different from how this was expressed in the history of writing. Or more precisely even: in two ways the movie *Metropolis* introduces us to the in/visible city that could not exist in the novel. This city came into existence because of cinema.

First of all, in the movie, it was not so much the people, but the massive skyscrapers that seemed to form the most important population of the newest cities (mainly in the new world, of course). This is obvious if we analyse how movies like *Metropolis*, and the many movies that practice this theme afterwards, work with their chiaroscuro. Chiaroscuro, the Italian Renaissance technique of painting with shadows, is translated in how the moving image practices appearance and disappearance with light and shadow. And this time, it is not the human face but the skyscraper, or better, the cluster of skyscrapers and the way the cameras and the sun (or an artificial source of light) relate to them, that matters: "Shadows of houses pursue the man running along the street" (Deleuze [1983] 1986, 51). Chiaroscuro was of crucial importance to the affordances of humanism since the renaissance, and now, with cinema highlighting this new urban space, its character and its style became of crucial importance to the affordances of the (new) city and the kind of life it allowed.

Secondly, much more so than in the novel *Metropolis*, in which the idea of the city was translated into existentialist and spiritual reflections of the protagonist, the movie explores the technological environments of humanity. In the movie, this new (shady) urban space seemed to ask for all sorts of (future) information, communication and transportation technologies. Unknown and unforeseen, all of these technologies were in many ways alienating us from each other. Through technology, the movie portrays a society subjected to objects of technology, to its powers and their consequences. Powers that were unimaginable to us before. This is the part we did know: that technology would somehow always benefit the powerful (or those who owned the technology and knew how to make use of it, for their own benefit). Technologies are not primarily tools to redact one's environment. Technologies are ways of sharpening the hierarchies (of capitalism) and of gaining social and political control.

In the movie *Metropolis*, technology plays a key role in the opposition between the powerless and the powerful. The high-rise apartments of the upper class, enlightened, spacious, clean and with tall windows, form a sharp contrast with the overcrowded, dark and dirty alleys of the underworld. This is the same opposition, as between the 'workers' and the 'thinkers'. Whereas with the first image the distance is realized by architecture, the intervention of the robots and by what was called 'the M-Machine' was needed to secure the second. Of course, in the movies that would continue this line of thought, this relationship between technology and power became more and more complex, and important, for both cinema and the city.

We need to keep in mind though that in our times cinema works differently from the cinema in the Interbellum and plays a different role in the everyday life of the city. The movie theatre (or the bioscope, to use its old and (etymologically) much more interesting name) is, on the one hand, no longer a dominant force in the streets, in the city centres of the world, as today, if these theatres are even 'in' the city, they have often moved to its outskirts, and are hidden within big multiplexes. On the other hand, the projection is no longer the analogue reflection of the city lights, but much more its digital imitation (the digital fakes reality, it is by all means a simulation as D.N. Rodowick so nicely puts (2007)). Cinema expresses symbolic information; it is not indexical anymore.

Nonetheless, although cinema itself may have transformed beyond recognition, movement-vision, which is that which produces our shared cinematographic eye, still functions in a similar way. Movement-vision still produces the vanishing point; an imaginary black hole located at the horizon into which all diagonals seem to disappear. It still desires a single light source, that is the sun, which organizes colour, produces illumination and shade, according to which the eye and the spectacle move. And lastly, mirroring the vanishing point, it is within the cinematographic eye (of the beholder), the focal point, where entire presentation, where the resonating lines and colours scattered around the picture turn into a fixed rhythm, come together and form the scene. In other words, the way cinema practices subjectification has not changed at all. In fact, with the further rise of the city and its new forms of control, the cinematographic eye seems to have become only more important.

What matters to analysis is therefore not so much the type of sign that cinema produces (the way in which the image re/presents reality). Rather, what matters is how the moving image resonates with the contemporary, how the rhythms of visual culture and the city respond to one another, mimic one another, develop, in relation to each other. One could say that cinema, as it came into existence in the 1920s, has been lost for good, since it is not related to 'space' in a similar way (movies are produced on the computer (and not in the city) and consumed in the suburbs, and perhaps even primarily, on the mobile phone). The cinematographic eye, on the other hand, needed the event as a starting point (the city during the interbellum), which, since then, became crucial in the realization of the kino-city, as Kochhar-Lindgren calls it, or the in/visible city as I referred to it. The rise of this new city in the end means that we entered an age in which the dominant forces of change turned visual/material, as he concludes (2020, 112):

> Screenings now organize the space of the city as the spheres of entertainment, surveillance and law enforcement, the economics of banking and shopping, the politics of social movements and elections, and the multiple modes of transportation that are folded together into overlapping digital networks.

Cinema needed the event to come into existence, and to persevere in being, similar to the in/visible city itself.

An Idea in Cinema

The in/visible city is in many ways an idea in cinema. Notwithstanding the way in which novels have explored this theme from Dostoyevsky to Zamyatin to the current writings of authors like China Miéville (think of his novel the City and the City (2009) in which two cities occupy the same space at the same time), or even questioning the reality of literature's cities in the contemporary even. All the cities that literature proposed to us already exist and come into existence over and over again; the 'other sides' they portray will continue to step forth from, and resonate with, the urban realities that we are engaged in today. Yet, the urban narratives explored by the movement image are in many more ways resonating with the urban spheres today.

Perhaps this is because, notwithstanding the existentialist and spiritual questions that contemporary city life is posing us, the megalopolises that dominate the world today (the clusters of cities that can often be found at the bigger deltas of the world, and rarely in the 'former West') are more than ever dominated by the dehumanizing BIGNESS (as Rem Koolhaas once called it in his manifesto Bigness, or the problem of Large (in Koolhaas a.o. 1995)) of their construction work rather than by the people that live there. The web of faceless skyscrapers masters the light and darkness (day and night), and the workers (the 99% as we now tend to call them) live according to them. Also, the communication technologies that traverse these spheres, not so much labelled 'robots' or M-machines but known—more abstractly—as algorithms, are the other dominant players of today. Buildings are the body of the megalopolis today, the algorithms of communication technology, their soul. Much like Fritz Lang and Tea von Harbou explored this with their movie on the in/visibility of Metropolis.

Perhaps this is also the case because the moving image, more so than literature, was able to show us, since the 1920s, how the city was being pulled apart, that those in power (the 1%) lived a completely different life compared to the 99%. When Karatani rereads Marx and Hobbes, he tells us that the kind of governance we associate with the state—but which he considers fundamental to all 'advanced' sedentary social organizations, to any kind of 'social contract' in the Hobbesian sense if you will—is never realized within a single community, and "that this kind of sovereign is not born from within the community through a process of self alienation, but rather originally comes from the outside—in other words, that the sovereign arrives as a conqueror" (2014, 69). This is very much in line with how Deleuze and Guattari ([1972] 1984), in Anti-Oedipus, talk about the merchant and the tradesman in the 'the primitive system' as they called it (within which the state/capitalism existed but was prevented to surface), whereas in the societies that followed, the merchant and the tradesman (and especially the network they were a part of) became the dominant form of social organization; in other words, "the death of the primitive system always comes from without" ([1977] 1983, 195). The rise of the city, which is the rise of a new capitalist sovereignty, is expressed through the cinematographic, as it is the moving image which shows us best how this new sovereignty arrives (necessarily) as a conqueror; aiming to redefine all that matters according to a new spatial regime, making the politics of visibility central to its idea.

Let me be very precise on what it means to 'have an idea in cinema'. In a lecture given at the FEMIS film school on 17 March 1987, Gilles Deleuze starts by saying that having an idea in cinema is a rare thing, just as having an idea in philosophy or in any other field doesn't happen very often. Henri Bergson, in the beginning of the twentieth century, had already said that philosophers, no matter how many writings they produce, had only one key idea, one main principle that was their guidance throughout their career. Deleuze seems to follow this line of thought, would perhaps say that 'having an idea' is even more rare. To have an idea in something happens only several times in a generation.

There are two very important thoughts connected to how Deleuze talks about ideas, which should be explored here. Firstly, there is Deleuze's insistence (we may call this his materialist insistence) that ideas cannot but happen 'in something'. As he puts it (2006, 312, italics in original):

> Ideas have to be treated like potentials already engaged in one mode of expression or another and inseparable from the mode of expression, such that I cannot say I have an idea in general.

This 'engagement' Deleuze refers to means that the in/visible city is inseparable from the 'mode of expression' called cinema; that the in/visible city expressed itself in many (different and profound) ways in cinema. The in/visible city happens in the moving images of skyscrapers, in the technologies that surround them, and in the people living according to them, as the analysis of *Metropolis* (the movie) above already shows.

The second thought that I find particularly useful in Deleuze's lecture, and that has not received enough attention in what has been discussed (but that will change), is the thought that such an idea in cinema realizes itself, in the moving image, through a 'cycle'. He introduces this term as such (idem, 319/20):

> a cycle that suddenly makes cinema resonate with the qualitative physics of the elements. It produces a kind of transformation, a vast circulation of elements in cinema starting with air earth, water and fire.

To claim that the in/visible city is an idea in cinema deeply engaged with the current city, then it starts by acknowledging that such a 'cycle' cannot be found in literature, in theatre or in other artforms of importance before. I started this chapter by showing that in literature there have been many writers who have an eye for the other side of the city, the dark side of town. And of course, these interests have also surfaced in other art forms. Of course, looking at how the Futurists—already in the 1910s—imagined the cities to come, the high-rise buildings combined with rapid transportation systems (as we saw them in the drawings of Antonio Sant'Elia for instance), may remind us of the cities that cinema presented them to us later. But let us not forget that for the Futurists, especially as they got more fascinated by fascism, cities were an ideal, something to strive for. For them, these modern cities (that seem to miss out on people, that seem not interested in the human scale, in human interaction even) were not only technologically but also psychologically and sociologically a major improvement to life.

In cinema, however, almost since the industry properly took off in the last years of the nineteenth century, this kind of optimism seems pretty much impossible. In cinema, the city, with its many different hierarchies (visibly and invisibly), almost has to imagine underground narratives of hopelessness and despair, as the numerous films that somehow tried to capture the radically new and 'inhuman' realities of urban life tell us. The coldness of the concrete, the images of walls spreading darkness, the ongoing technical procedures, that take up so much time; haunted by the high shadows of buildings, the in/visible city leaves little room for reflection, according to the movement image. New York City (a.k.a. Gotham City) served as its favourite example. Since the 1970s, Hong Kong appeared as its oriental double.

Emphasizing the importance of the 'cycle', means that, when we talk about the in/visible city being an idea in cinema, we need to stress that this idea makes both the materialities of cinema and of the city (the megalopolis) to resonate with the narrative on screen. The cycle is thus not so much 'limited to cinema', and actually I guess that the cycles of all ideas necessarily work in at least two modes of expression. In this case it runs through the elements of cinema, but also through the elements of the city. Like a screw, the cycle turns and twists, and connects the different modes of expression, makes them resonate together, makes them belong together, cinema and the city, this time. Deleuze already told us that such 'engagements' are not random, they are a necessity. An idea like in/visibility is thus never simply 'thought of', meaning it is neither a product of the human mind; it is not the I-think (the Cartesian perspective) neither a critique (the Kantian perspective) in response to city life. On the contrary, the cycles of the in/visible city are "given rise to", as Deleuze would put it elsewhere (280), by cinema and the city together. Great filmmakers could only have picked up this idea in relation to how both the matters of cinema and the matters of the megalopolis resonated together. Different in/visibilities were distilled from the vital fabric it functioned in/anticipated upon.

Joker

Todd Phillips' movie *Joker* (2019) offers us an in/visible city, more or less imagined in New York (or more specifically, in the South Bronx) in 1981, the place where he himself (like many of his crew members) grew up. A city which was dominated by worn out, red-brick and concrete high-rise buildings, garbage and waste, sets the stage for an environment that is poor in every way, offering no perspective, no future, to any of its dwellers. The light is dimmed, the rain keeps falling from the sky, there are no trees and hardly any birds in the city. In most of the scenes, when the protagonist returns home late from work, the streets are desolated, as if even the humans have retreated from public space.

Philips situated Arthur Fleck in this environment in a tiny rental apartment with his mentally ill mother. It is an environment in every way different from the place where his alleged father, Thomas Wayne is said to live. Thomas Wayne, (also) the father of Bruce Wayne, a young boy still in this film but later also known as Batman, is extremely rich, chairperson of Wayne enterprises, a multinational company also in charge of the city's most important media outlets. Wayne enterprises is situated in Wayne tower, and Thomas has already announced running for mayor of Gotham.

Whereas Thomas Wayne is 'connected' to life in so many ways (technologically, socially, economically of course), the life of Arthur Fleck, the person about to transform into Joker, and of his mother, is part of a city where alienation, in the post-industrial sense, is key to how people live their lives. Alienation determines how the 99% relate to the (run down) built environment, the outdated and hardly functioning technology, to the absence of companions with whom we share this earth, including, to our fellow humans. Obviously, this is reflected in Arthur Fleck, the main character of the film, present in almost every scene. Alienation dominates the film, as the individual scenes show us what the forces of capitalism, the powers of Modernity, have accomplished by now; Fleck, the protagonist, supposedly surrounded by others (living in the heart of the city), has to live his life alone, detached from everything around him. Everything that surrounds him is not sympathetic to him, is not in touch with him. Arthur Fleck is invisible.

Invisibility, in so many ways, has turned into the major wound of late twentieth-century (and early twenty-first-century) 'modern times'; both the lives on-screen (in cinema, in all the new mobile and public screens) and the lives off-screen (our neighbours in the condensed megacities that are now populated by 60% of the world's population) take place in solitude. Screamed at by adds, by the masses in the subway, by the fellow travellers with whom we all flow through the interconnected networks of imagined and urban life (if there is any difference between the two at all), one moves alone. And this is not the fate of only 'the lowest' of classes in society, not just the untouchables or the working class; it is the life of the 99%. Especially today, all of us are invisible; all of us live in solitude.

Except, of course, to those in power of course. Big tech firms and states have a very clear vision of every single city dweller on this earth. Practicing intensive human farming, where the subjects are continuously ear tagging themselves with their mobile phones (paid for by themselves too), all of us are visible in every step that we take.

Perhaps I should refer to this as 'the Marxist moment' in their plots: movies that are somehow playing with the idea of the in/visible city; all seem to be in search of a possible revolution, for a way to turn around the in/visible organization. This revolution is always sincere and needed, and is, in contemporary cinema, all too often embodied by the radical unworldliness of a deeply complex and (often) deeply sadist protagonist. A complex figure it is, as this aesthetic figure necessarily embodies a series of 'layers' (histories, conditions, phantasms) that are twisted and cracked, that are more or less mirroring and incorporating the 'idea' of the city they find themselves in. Embodying the in/visible city, the figure has to be sadist, ready to destroy its loved ones. The twisted and cracked layers of history that come together in this figure always have a very strong desire to express themselves by means of a series of highly aggressive physical attacks, directed at the status quo, at normality. A scattered and scattering persona, Joker, as he completes the last stage of his metamorphosis, and acts and looks like the character we remember from the comic books, seems to have switched positions with the 1%, but not by taking over the power to keep the status quo. On the contrary, still the underground man, its aim is to finally dismantle society, to dismantle all of its inhabitants. Always happy to feed our deepest, darkest and most destructive desires, Joker, with a smile, not so much 'introduces' evil to the city, but rather unveils the frustrations and darkest desires that were essential to the in/visible city in the first place.

Of course, the smile is crucial here. The smile is often symbolizing the Joker's madness, his abnormality, and places him in sharp contrast with the Batman, the saint, he who wants to restore normality, and who has never been caught smiling. This time, however, is it the smile which grants Joker his visibility. Like a Cheshire cat, it is the smile that remains.

It is remarkable that, although with little knowledge of what caused the smile, we, the spectators, have previously always been convinced of the presumed madness of Joker, even when he was played by Heath Ledger (in the *Dark Night* trilogy), whose performance already gave us a Joker that actually could have been one of us. Still, we were not supposed to sympathize too much with this figure. On the contrary, the response to Joker has always been that we were longing for the moment the horror was over (near the end of the movie), we were longing to see Gotham City return to normality, the city as it was 'before' the revolution. The Batman, the invisible disguise of the super wealthy Bruce Wayne, fortunately, restored normalcy. Joker was of course not so much defeated, but at least he was invisible again. Taking off his mask (/paint) Joker left the scene, the bat could return to its cave. Alas, Bruce Wayne and his companions were visible again, in control of Gotham City. Normalcy had returned.

So many histories come together in this idea. Growing to fame in the 1920s, the comic books that gave rise to the superheroes were already working with an urban dystopia, though one could argue that for them, it was—more simply—the dysfunctional city that they were struggling with. Perhaps we should say that this was even the era in which the dysfunctional city was seen as the dysfunctional society, as the epicentre of the many wounds that had surfaced in European and American cities after World War I and that would necessarily lead to World War II. In the 1920s, too much happened at the same time, in terms of economics, in terms of art, but of course also in terms of politics (the rise of Fascism) and urbanization. A traumatizing time, in many ways, and perhaps only the superheroes (the abnormal, not to be confused with the 99%) were considered capable of bringing bring back the normal again. In different ways, in different times, they kept on doing so ever since. Todd Phillips' 2019 version of *Joker* might be interesting as a 'character study' of how an extremely violent and narcissistic villain comes to be (as the director himself has put it in interviews), it is even more interesting as a cinematographic tale of late capitalism, in which the idea of in/visibility is expressed through the invisibility of Arthur Fleck and the 99%, and the visibility of the smile.

How telling it is that the violence and the destruction, which had always been part of the city, which had been destroying the lives of many since long, are only caught on camera in relation to the smile? Note that Joker started his killing spree only after he was harassed by three young men, 'coincidentally' Wayne Enterprises businessmen, whom he shot as an act of self-defence. Perhaps it was only with the last character, whom he killed while he was already wounded and laying on the floor defenceless, that Joker took control, replacing Arthur Fleck. As expected, the killings are condemned by Thomas Wayne, who, remembering that the victims were his employees, immediately concluded that all those jealous of his success (of the success of the people in power, the 1%) were clowns. The newspapers, owned by Wayne, immediately, classically, copied this narrative. The people, however, sympathize with Joker, who had always been one of them. They start wearing Joker masks to reveal their sympathy, to make visible that their lives too need to be seen.

"We are all clowns", it is said (and written, on banners during protests) several times near the end of the movie. Not because the people envy 'the success' of Wayne, but because putting on a clown's face makes them visible to their conquerors (the ones in power). Putting on the mask that smiles reveals how their lives, their cities have been destroyed. The mask that smiles expresses the idea that the people and the city share a bad childhood, that remained unnoticed, indeed invisible. Unable to express themselves, since the technology was in the hands of Wayne enterprises.

The cinematographic cycle of in/visibility may have been set in 1981, but in every way 'takes place' in 2019, the year of its release. In so many ways, this cycle connects to city life today, to the late capitalist struggles of today, to the invisibility of those with no power at the start of the twenty-first century, that the idea in cinema it expresses demands a reading of the contemporary.

Just before its release, on 26 September 2019, NBC news reported that movie theatres in America banned Joker costumes in their theatres, following a warning from the US Army. Movie chain AMC banned face masks, make-up or anything that would 'conceal' the face, while Landmark Theatres forbade costumes of any kind. The US Army called Joker a 'potential threat' and spread memos with a warning for a mass shooting in movie theatres, inspired by the violence in the movie. The US Army was following an FBI warning even, in which the bureau signalled the popularity of the Joker character with some seemingly violent internet groups. Given all the publicity that followed and the way it sparked fear with the industry, and the fact that the movie was therefore instantly condemned for its violence, for its 'dangerous message' and for the overall 'nihilism' that marked its main character, it is quite a miracle that in the end the movie turned out to be such a global success. Financially the movie was a huge money maker with a budget of around sixty million dollars, it raised over one billion dollars with the box offices, making it (and this is very interesting) the highest-grossing R-rated film of all time. In terms of accolades, it was especially the performance by Joaquin Phoenix which gained the film professional recognition, though also the music score (made by the cellist Hildur Guðnadóttir) received much critical acclaim.

Hong Kong

> Whether it's good or bad, it is sometimes very pleasant, too, to smash things. (Fyodor Dostoevsky, *Notes from the Underground*, 31)

I was living in Hong Kong the moment I first saw an announcement of *Joker*, in May 2019 already. Hong Kong, the former British colony, 'returned' to China after a hundred-year lease in 1997, saw its rise in the 1970s and 1980s. An unprecedented rise that, as always, happened on many levels. Most obviously, the number of skyscrapers that dominated the city grew dramatically; Hong Kong is still the city with by far the highest number of skyscrapers in the world, with 480 buildings higher than 150 metres, almost doubling New York. Of course, the economy skyrocketed; Hong Kong has since long considered itself the freest economy in the world, and, for the past half a century, it became one of the world's most significant financial centres and commercial ports. Hong Kong also quickly became a global centre for technology and media for East Asia and of course became one of the world's most iconic centres for contemporary cinema, with a thriving industry of itself (with Bruce Lee as a major accelerator in its early days) but also as an iconic image in practically every movie in which the East, the 'Global city', or simply the contemporary urban sphere, had to be expressed.

Regarding its wealth, one final thing needs to be added here: being the freest economy in the world really means that its rulers implemented a quite radical 'laissez-faire' style of capitalism, inspired by Milton Friedman and the Chicago School of Economics. This among others has led to the fact that after its acceleration in the 1970s and 1980s, Hong Kong has by now turned into the most expensive city in the world, especially in terms of real estate, and is known for its high level of income disparity. Housing a large number of millionaires and billionaires (only New York is said to house more of them, which can be questioned since these days, an almost equal amount of wealth is 'officially' living in neighbouring Shenzhen, and the borders are not waterproof), a significant part of its population lives in poverty, in the tiniest apartments (sometimes no more than 'caged beds'), working long hours under bad conditions.

As with many colonial governments, the Hong Kong government (in colonial times and in the current situation) has hardly implemented any trade control. Termed 'positive non-interventionism' by its financial secretary John Cowperthwaite in 1971, Hong Kong, instead, has been under control of the global forces of capitalism since long. So many of its transformations, as they were realized over the past fifty years, were symptomatic for how capitalism imagines the city, its architecture, its technology, to develop. It makes perfect sense to consider Hong Kong, more even New York, a deeply cinematographic city; to consider the city so very much entangled with the global cinematographic cycle, that the city shapes and is shaped by the movement image. Its neon signs, the Kowloon Walled city with its maze-like street plan controlled by triads, its martial art fighters and street prostitutes, have been replaced by extremely expensive apartments, white-collar (fin-tech) crime and its vertical organization, but its signature is Hong Kong all over. Of course, especially the idea of the darkness is embodied by it.

With unprecedented speed, Hong Kong, over the past decades, has become the model for the contemporary 'glass city', very much according to Zamyatin's ideals; a transparent city run by OneState, known as capital in this case (though the difference between capital and data is practically inexistent these days). Zamyatin's emphasis on numbers (as said, everything is already datafied in his 1920's novel), on totalitarian rule and on seeing the city as a technological organism is still the dream of Communist China it seems, 100 years after Zamyatin saw this happening in the Soviet Union. This way, Chinese cities have already become great examples of how the control societies, that Deleuze predicted in his famous 1990's essay, have been realized. Many of its major cities (including Shenzhen, the city bordering Hong Kong) seem to at least strive to become societies in which all forms of disciplining (the prison system, the school system, the hospital system, the business system) have merged into one entangled digital system of domination. In Hong Kong, it is (still) capitalism which rules, subjected to the ideals of the 'free market', an equally complex and hierarchical system of organization (as Karatani showed us) but in many ways very different from how the state functions. In fact, over the last decades, Hong Kong is probably the best example of an urban centre made of high-rise glass constructions and data flows. More than any other city in the world, cinema therefore sees the city of Hong Kong (and especially its top floors!) as the archetypical city of control.

Of course, being so much a part of the global arena of capital, Hong Kong cinemas had been the home of many Marvell/Disney superhero narratives in recent years, many of them at least partly shot in this city (or at least pretending to be so). The online trailer of *Joker* was intriguing as it immediately broke with its typical narratives. As we all know, superhero narratives voice the longing for a return to normality, for the defeat of madness; for society's 'strong arm'—in whatever fantastic form—to regain control. And although these films often take place in the major cities of the world, their 'idea' does not so much resonate with the undergrounds (as Dostoyevsky would call them) that have firmly nested themselves in the cities of today. The architecture and the technology are superficially present in these movies. But none of these movies seems to worry about the lives that populate these cities. Practicing an almost fascist desire to return to normality at any cost (not uncommonly, buildings and lives (of the common people) are destroyed in exchange for regaining the status quo without any regret), none of these movies is engaged with the in/visibility of these cities as conceptualized above. The cinematographic lessons learned from *Metropolis* and movies alike are superficially reflected in the chiaroscuro of the skyscraper city, and the camerawork that comes with it. In sum, your average superhero seems to be interested only in restoring normality (the rule of capitalism and the state, and their in/visible strategies) in Hong Kong or New York. The trailers from *Joker* showed us something very different.

October 2019

> No hesitations, no delusions. There is only one truth, and only one true way; this truth is two times two, and the true way-four. And would it not be an absurdity if these happily, ideally multiplied twos began to think of some nonsensical freedom—that is clearly, to error? (Yevgeny Zamyatin, *We*, 67)

Unfortunately, after seeing the trailers online, I found out the release date for *Joker* was set for October that year. After spending the summer in Europe, I did return to Hong Kong in October. However, while I was gone, Hong Kong appeared to have turned into a very different city. In the years after the handover in 1997, I had spent more and more time here, and the difficult relation between Hong Kong society, as it was formed under capitalism (under colonial 'rule'), and the new Hong Kong leadership, in ever closer cooperation with the authorities from mainland China, turned out an increasingly problematic issue. The Hong Kong communities had always been ruled by outside forces and when their British rulers were exchanged by rulers from Beijing, they were never even consulted (the handover itself was a deal set up in the 1980s by Margaret Thatcher and Deng Xiaoping and did not involve the voice of the people from Hong Kong). Looking down upon the city, from their multi-million-dollar apartments, the ruling class, made up of capitalists before (with a strong input of the British financial system which connects Hong Kong to New York, Singapore and London and which still rules the world), is slowly being replaced by statesmen from mainland China. And as these rulers, still, more than anywhere in the world, live a life completely different from how the 99% struggles its way through, Hong Kong offers the best possible mis-en-scene for a current day Metropolis.

It is not surprising that the protests that had always been key to this territory, because of its strong rule, and strong hierarchy, have intensified since the handover (especially since the 2014 Occupy Movement in Hong Kong, also known as the Yellow Umbrella Revolution). Starting in May 2019, however, when more and more people got sceptical of a bill would have allowed extradition to jurisdictions with which Hong Kong did not have extradition agreements (including mainland China), a renewed protest movement grew rapidly, which lead, among others, to a peaceful protest at Victoria Park attended by two million people (which is one third of the total population of Hong Kong) on 16 June 2019. But as results stayed out (the law was abandoned after a while, but the other demands of the protesters were never met), the protests were more and more dominated by frustration and disappointment.

On 20 October, I was in Mong Kok and Tsim Tsa Tsui, on the Kowloon side of Hong Kong, roaming the streets of a city under siege, when police and protesters played cat and mouse. Gone were the days in which the general public went out for peaceful protests, in which they jointly asked their leaders to change the course. Of course they had long felt that things were heading in the wrong direction, that they were heading for a catastrophe, probably since the signing of the Sino-British joint declaration on 19 December 1984. Since May, they turned to the Hong Kong SAR Government, and more in specific, to Carrie Lam, the chief executive of Hong Kong. Who else to turn to? Perhaps their five demands were unreasonable/impossible to meet, but there simply was no road that would lead them to the rulers, there was no way to get inside of the impregnable fortress, as Kafka would have put it.

How else can one express a deeply felt fear, that a disaster is upcoming? In *Joker*, since the start of the movie, Arthur also felt that things were about to go very wrong. He tried everything to prevent it. So, he went to his (nameless) therapist/social worker, played by Sarah Washington, and he complained about the fact that his medication wasn't able to let him live according to the cause of things. Of course, medication could never solve the problems, it was only meant to suppress (or delay) the consequences. And after higher powers decided to stop the counselling and the medication, Washington tried to explain to Arthur that she was not able to get him inside the fortress, and concluded: "listen, nobody gives a shit about us. They don't care about you and they don't care about me either."

It is not easy to see how Carrie Lam could have said anything different to the Hong Kongers. In her media performances she tries to figure as Maria, the character from the movie *Metropolis*, the maternal figure from the working class, noble and idealist, who is able to unite the haves and the have-nots. In the eyes of the protesters, however, she became

Robot-Maria, the artificial double, a sinful and chaotic entity, tasked with leading the change towards the mass destruction of the city.

It was obvious that at the end of October 2019, after the people of Hong Kong found out there was no way to get inside the fortress, that their therapy and medication had stopped and that, despite all their efforts, they would remain invisible to the outside powers that controlled them. Of course, the protesters got bitter, more violent and very frustrated.

Masks

And so, the people of Hong Kong put on their masks. They covered their faces, dressed in black, put on big plastic glasses and sometimes added a yellow construction helmet, in order to become visible, to unite as a group, to speak truth to power. Sometimes they completed their gear with an umbrella, preferably yellow, as a reminder of the 2014 'yellow umbrella revolution', Hong Kong's part of the global occupy movement, in which they, kickstarted by other legislative developments but by the same histories of pain, also protested (and occupied key areas in Admiralty mainly for 77 days). Photo number 2 shows 'Lady Liberty', a statue created by the protesters, dressed accordingly. It was placed at Lion Rock (a rock in Kowloon which offers a view of the city, considered to represent the spirit of the people of Hong Kong) for one night in October 2019 (Fig. 10.2).

The response from control was telling; on 4 October 2019, Chief Executive Carrie Lam invoked the Emergency Regulations Ordinance to impose a law to ban wearing face masks in public gatherings. Confronted with ever more violence, this move could be interpreted as an attempt to curb the ongoing protests, but in every way, its aim was to make those communities searching for a way to be seen, unrecognizable (invisible) as a group, and easily recognizable (visible) for the rulers. Back to square one. Similar to how Thomas Wayne projected all the violence upon the clown, the rulers of Hong Kong projected all the violence upon the single 'rioter', dismissing the long history of violence that preceded it, that lead to this explosion and that would tell the real problem (which was complex, and 'invisible').

Fig. 10.2 Lady Liberty at Lion Rock, Hong Kong, photo by the author

The protests I witnessed on 20 October seemed by all means a highly ritualized scene. A very significant one though. It was a 'regular' protest, the kind of protest that would take place almost every evening back then, and one which, before the corona virus dominated all of our lives, could also take place in many places around the world (from Barcelona to Santiago de Chile, from La Paz to Beirut to Delhi (Jamia Millia Islamia University)). There was a spectre haunting the world in 2019, its presence was caused by many different histories, but everywhere, somehow, the community felt under threat of its outside rulers (capitalism, the state), everywhere the oppression became too much, and caused uproar, violence. Everywhere, city life, somehow, became impossible, and needed to be given a face, needed to be seen. It was no coincidence that the mask of Joker was spotted at many of these protests, in the Halloween celebrations (or local variations upon this theme), at posters, or via stickers.

Somehow, the idea that "we are all clowns" matched the state of the city everywhere. Somehow, the ruins of colonialism, the expanding hierarchies of capitalism, made life impossible. The warning signs were not picked up, an explosion had to follow.

Hong Kong, however, should in many ways be seen as the starting point of these unrests. Or at least as its fulcrum, as so many of the wounds of the world came together here in Hong Kong. With one third of its population hitting the streets, what happened, with a bitter undertone, could be referred to as the choreography of our time.

On that day, a large group of people from all generations was walking down Nathan Road, singing and chanting, up until the moment riot police arrived. This was the sign for the 'civil guard' to step forward. Whereas Joker protected himself with a smile, the black-clad youngsters protected themselves with (yellow) umbrellas and facial masks against heavily armed combat soldiers, many of them were wearing CCTV cameras, had stroboscopes on their helmets (in order to confuse their 'opponents'), and were covering their faces with gas masks. Roadblocks, assembled from stones, building material and fences as found with nearby construction sites, were built to stop the one, water cannons and tear gas, produced in high-tech arms factories across the world, were used to disperse the other. Out of nowhere, the international press and the first aid volunteers rushed to the scene, gathering at the no-man's-land, where the two sworn enemies would never meet, like empires at a boxing match that never commenced. Here, and only here, the images were made that filled the news broadcasts all over the world for months; this was the twenty-first-century warzone, where so many of the wounds from the past 200 years were opened up; reminiscences of colonial pasts, the rise of a twenty-first-century global power, or, more generally, the ongoing struggle for control.

I was standing, with my partner, together with the press and some other 'onlookers', at less than a metre from the scene, on the pavement. Protesters warned us to be careful, perhaps because we were wearing black like them. Overseeing the scene, we saw the protesters pulling back, dispersing across the main street and some backstreets. The front line was only visible because of how the police operated; in large well-organized legions, they marched forward, clearing the streets. Through a megaphone, in Cantonese, they were ordering the protesters (who had already left the scene) to "stop the violence". Next to the policeman shouting, there was some graffiti of a gun and the words "stop shooting people" and "stop tyranny". Before, the protesters had vandalized stores with public sympathy to mainland China, and the entrances to the MTR stations (following the orders from the SAR Government to stop their services during protests, the company was seen by protesters as a collaborator). Afterwards, facial recognition technology would allow the police to identify perhaps some of the protesters, charge them with rioting, taking part in an illegal assembly, and prosecute them. At this hour, however, there was hardly any violence, from neither side. One protester got tear gas in her eyes and was taken care of by volunteers from the Red Cross.

Joker was released in Hong Kong on 3 October 2019. On 5th at midnight, the anti-mask law went into effect, forbidding any form of face covering (including paint) at a lawful rally or march, unlawful or unauthorized assembly or during a riot. At the protest we visited, the iconic phrase of the movie "we are all clowns" was sprayed with graffiti on the roadblocks and flyovers. Posters and stickers featuring Joaquin Phoenix as Joker, accompanied with the sentence "Mask on, Hong Kong 2019", were to be found throughout town.

The idea of in/visibility, as imagined in cinema and in the post-1920 city, disrupted normality in Gotham City as it disrupted normality in Hong Kong. And this was not just because the protesters copied the face of Joker but because they sensed a common ground. On the contrary, what happened in Gotham City and in Hong Kong, happened simultaneously, by which I mean that the 'cycle', as Deleuze called it, of the idea of the in/visible city, mattered to Gotham City while it mattered in Hong Kong.

The next day, the sun rose for the happy few in the top floors of the skyscrapers and not for the 99% that lived 'underground'.

Towards a Technology of Affect

Deleuze said "The act of resistance has two faces. It is human and it is also the act of art" (2006, 324). This is obviously reflected in *Joker* as a figure, fabricated as a consequence of many histories of violence and especially of how 'the smile' had made his existence since childhood possible and impossible. But of course, all of the make-up, the green hair, the colourful suit, they were all necessary in order for resistance to be realized, to stand out, to be seen. But the creative intervention that was realized with Joker of course included the whole of Gotham City, all of the masks did not cover up the sad face, but uncovered the anger of the people, the frustration. In the end, even, all the violence, the burning of cars and the smashing of the windows, as it happened in the rage when finally, things exploded, uncovered the anger of the people. Contrary to the technologies owned by Thomas Wayne (think of all the media firms that were part of Wayne Enterprises), Joker and the 99% aimed to take back their city through what should be called 'technology of affect'. Grounded in challenging the in/visible, creatively reimagining the objects within reach, a resistance of the underground was realized in a final, ultimate, attempt of the community to question the power of the sovereign. Clearly connecting the deeds of the Hong Kong SAR Government to the will of Beijing, it was a surprise to many that the protesters more and more turned against the forces of capitalism too. But of course this made sense, as especially in rage, blind of fury, the clusters of powers clustering all these histories that caused this impossible situation had to be pushed out of the city altogether. This was the moment that all the sovereigns had to go.

More than ever (in the globalized world of today), control comes from elsewhere, from outside the community, and the means to realize control, especially now, have little to do with what is available in the streets anymore. The facial recognition cameras, the high-tech gear worn by the police, billions of dollars have been spent on new technologies, new data systems that are all developed in far-away expensive military labs are at the service of those in power and are employed with the sole goal of establishing a system that is able to gain back the control over the city as soon as possible. The protesters on the other hand prove themselves very much a grassroots movement with their use of everyday utensils like umbrellas, laser pens, yellow helmets and black clothing (and occasionally with make-up and colourful suits). The use of traffic cones to extinguish tear gas canisters, carpets of bricks to stop vehicles and all sort of material from construction sites to stop, distract or attack the riot police are all perfect examples of a technology of affect. It is a creative and artistic way of occupying the streets you love, occupying them by heart, by will and by chance, in order to gain back control over them. Never designed as weapons, as means to defend one's ground, all of these objects are discovered anew. The rage, therefore, is not about creating chaos at all. On the contrary, this creative act is always a battle against chaos. Destroying the streets, and even destroying the shops of one's city, can also be done out of love, which we know since antiquity; sometimes destruction is needed in order to facilitate new life, new more sustainable forms of living.

It asks for the utmost creativity, to find ways to turn a lyre into a bow.

In his famous text Postscript on control societies, Deleuze says that there is no need to fear the new integrated systems of free-floating control, nor should we hope for the best; we need to find 'new weapons', that allow communities to resist, to step up, become visible again, as a community. Bruce Lee's famous adagio, "be water, my friend", has often been memorized by the protesters as a strategy to deal with those in power. Artist Kacey Wong remembered Bruce Lee's words and concluded in his TEDxVienna talk "How can I become formless and shapeless in a protest? When you put people in the street, they have to become the street."

What Remains

When stores were damaged, they were quickly covered with wooden planks. The walls filled with graffiti were often painted over the next day. Hong Kong photographer Chris Gaul perhaps captured this best with his photo series called 'Erasure' where he, right after the cleaners (working on orders of the government), quickly erased the man-size characters that were placed on the glass plates of bus stops and similar locations just the night before. Gaul himself said in an interview:

> Even though most were unreadable, I was struck by how they capture the energy and essence of the moment in Hong Kong: upheaval, collision, anger and violence—but also resistance and hope.

Arthur notices that the complexities of life (the city and its others, the histories and their presences) have deformed him, made him unrecognizable to himself. Arthur was used to playing the clown. It was what his mother taught him to do; it was how he, as a grown-up, hoped he could mask his out-of-this-world-ness, his nervous laughter, his 'weirdness'.

Arthur tried hard to be a clown. Arthur made jokes that no one understood, like all of the wounded, tragically deformed by the powers of the father, the mother, by the powers that control. The moment he starts to live his wounds beautifully, is the moment he is not dancing alone anymore, invisible in a shady bathroom or in a shabby apartment, now he has a huge audience cheering for him.

But he couldn't persevere in being like this because of society. In 1947, French theatre maker Antonin Artaud, whose physique keeps reminding me of Joaquin Phoenix's *Joker*, always in a struggle with his mental condition, wrote a beautiful homage to Vincent van Gogh entitled *Van Gogh, the Man Suicided by Society* in which he stressed that van Gogh, also struggling, did not kill himself; society "had him punish himself" (2004, 1443) for how he dealt with his alleged 'madness', his weirdness, his out-of-this-world-ness. Artaud himself died of an overdose of chloral hydrate in a psychiatric clinic a year after he wrote this text.

The movie *Joker* ends in vain. It is unclear whether Joker will continue his life in a psychiatric clinic, in jail or anywhere else. It would not make sense, as it happens with so many of the movies on superheroes, that there would be a part 2 in which Joker returns, disturbing normality again, after which he would become invisible again. That is not how the visible and the invisible city work. That is not how revolution works.

In an interview, Deleuze once stated ([1990] 1995, 171):

> They say revolutions turn out badly. But they're constantly confusing two different things, the way revolutions turn out historically and people's revolutionary becomings. These relate to two different sets of people. Men's only hope lies in a revolutionary becoming: the only way of casting off their shame or responding to what is intolerable.

People's revolutionary becomings will never be understood by their conquerors if only because they cannot see what happens underground, they cannot see how a new community is already in the making, how new weapons are being forged. It is a mistake to think that the desires of Joker could even be fulfilled by his visibility. That is not how visibility works.

Wounds never heal. They transform and transpose until one day, they surface again, unforeseen and unrecognizably, to speak truth to power.

Put on a happy face.

References

Artaud, Antonin. 2004. *Oevres*. Paris: Editions Gallimard.

Deleuze, Gilles. [1983] 1986. *Cinema 1, the Movement Image*. London: The Athlone Press.

———. [1990] 1995. *Negotiations*. New York: Columbia University Press.

———. 2006. *Two Regimes of Madness. Texts and Interviews 1975–1995*. New York: Semiotex(e).

Deleuze, Gilles, and Félix Guattari. [1972] 1984. *Anti-Oedipus: Capitalism and Schizophrenia*. Minneapolis: University of Minnesota Press.

Dostoyevsky, Fyodor. [1864] 2018. *Notes from the Underground*. London: Global Grey Ebooks.

Harbou, Thea von. [1925] 1927. *Metropolis*. London: The Readers Library Publishing.

Joker. 2019. dir. Todd Phillips. Warner Bros.

Karatani, Kojin. 2014. *The Structure of World History*. Durham: Duke University Press.

Kochhar-Lindgren, Gray. 2020. *Urban Arabesques. Philosophy, Hong Kong, Transversality*. London: Rowman and Littlefield International.

Koolhaas, Rem a.o. 1995. *S,M,L,XL*. New York: Monacelli Press.

Metropolis. 1927. dir. Fritz Lang. Universum Film.

Miéville, China. 2009. *The City in the City*. London: Palgrave Macmillan.

Rodowick, D.N. 2007. *The Virtual Life of Film*. Cambridge, MA and London: Harvard University Press.

Serres, Michel, and Bruno Latour. [1990] 1995. *Conversations on Science, Culture and Time*. Ann Arbor: University of Michigan Press.

Zamyatin, Yevgeny. [1920] 1972. *We*. New York: Avon Books.

11

Techno-Medieval: Rise and Fall of Contemporary Metropolitan Networks

Emine Görgül

Introduction

The initial outline of this work was drafted in the midst of ISIS, Islamic State of Iraq, and Syria's invasion of the city Raqqa, and the circulation of drastic images of violence, not only against the humanity and the civil life, but also against the built environment. The purveyed images and news asserted the destruction of the thousand-years-old historical cities in and around the milieu, whereas the ongoing systematic operations loomed as a vicious cleansing of gradually accumulated tangible record of civilization. Significantly, the ancient Roman temples like Baal in Palmyra[1] (Fig. 11.1) and the historical city itself—where was once the

[1] Please see, A. Currey, "Here Are the Ancient Sites ISIS Has Damaged and Destroyed", *National Geographic*, September 1, 2015, Washington, DC: National Geographic Society. https://www.nationalgeographic.com/news/2015/07/150702-ISIS-Palmyra-destruction-salafism-sunni-shiite-sufi-Islamic-State/ Retrieved: 7 November 2020; K. Romey, "ISIS Destruction of Ancient Sites

E. Görgül (✉)
Istanbul Technical University, Istanbul, Turkey
e-mail: gorgule@itu.edu.tr

S. S. Das, A. R. Pratihar (eds.), *Technology, Urban Space and the Networked Community*, https://doi.org/10.1007/978-3-030-88809-1_11

Fig. 11.1 Baal Shemin Temple Blown Up, Palmyra, Syria, 2015. ([URL]: https://www.nbcnews.com/storyline/isis-terror/isis-releases-images-purportedly-showing-bombing-baal-shamin-temple-palmyra-n415496, Retrieved: 20th, 2021)

border station city of Roman territorial network and the essential terminals of the trade route—were destroyed irreversibly. Thus, the mediation of all these images paved the way for a further inquiry to define the *in situ* spatio-politics and to describe the impasses of the situation in the frame of contemporary urban condition and its criticality. Accordingly, the enquiry transformed into a research for tracing and theorizing the transitive processes, limits and crisis of the metropolitan condition for the contemporary society, with its decaying infrastructures under the influence of destructive forces.

Then again, these severe news and images further depicted a huge ideological contrast or a dilemma between two diverse mind-sets and established ethical values based on progressive rationalism on one side, and the radical destructive and inquisitional medieval gestures on the other side.

Hits Mostly Muslim Targets", *National Geographic*, July 2, 2015, Washington, DC: National Geographic Society. https://www.nationalgeographic.com/news/2015/07/150702-ISIS-Palmyra-destruction-salafism-sunni-shiite-sufi-Islamic-State/ Retrieved: 7 November 2020; M. Ahmed, "Why Does ISIS Destroy Historic Sites?" Centre on Religion & Geopolitics *Tony: Blair Faith Foundation Web Archive*, September 1, 2015, https://web.archive.org/web/20151006100612/http://tonyblairfaithfoundation.org/religion-geopolitics/commentaries/opinion/why-does-isis-destroy-historic-sites Retrieved: 7 November 2020.

In other words, the interrogations like the uncanny of why such centuries-long knowledge of urban network making and the infrastructures of the civilized society were discarded at the ultimate age of humanity, and how this dark middle-century manner of feudal-dogmatic perception was assaulting all achievements of enlightenment and civic-ness, while at the same time bizarre attitudes like taking over the advantages of latest technology of smart gadgets and drones, to visualize and circulate the bigotry legacy of these destructive groups and individuals, have been the triggering questions to commence this work in the search for an explanation about the *zeitgeist*.[2]

Nevertheless, neither a political nor a teleological criticism of contemporary religious fractions, as well as their radical practices acting as marginal sovereign powers inhabiting the milieu are the central discussion of this work. Conversely, on the basis of either political deployment of space and time, or the manipulative capacity of urban realm as a tool for ideological representation, the presence of urban condition and the related networks are essences of my debate. Yet, the decay of these components, or their interruption in spite of marginalizing the cities into deterritorialized entities of global terror and chaos, and the mentality of regression have been the principle indicators in developing this work. On the other hand, the irrational engagement of these radical groups with everyday technology and its hostile use for the show off in destructing their destructing capacities against the infrastructures and networks of everyday life and the civilization history have been the central concerns for theorizing a novel argument on the recent crisis of metropolitan condition and the related networks while questioning the rise and the fall of urban realm within the last hundred years of modernist agenda.

In this respect, the notion of "Techno-Medieval" is proposed as an invented term by the author, to start dwelling the critical stance in describing this complicated and intertwined even multi-fold relations of the current ecologies of mouldering spatio-politics, as well as mismatching mind-sets of regressive mentality and its technological

[2] In addition to these tandem observations, the author's criticality mainly emerges from the fact of her position being as a residence of a neighbouring country in the milieu, while being affected from the indirect reflections of the ongoing struggles via massive refugee wave and off-border politics of the country in the consequence of events.

interdependence. Yet, this invented concept is further utilized to encompass these numerous incoherencies of technologically equipped, and globally affecting radical violence of marginal groups or individual that have been behaving and acting in the mind-set of the middle age, or responding to the events in the way without utilizing the rational perspective. But rather, they seem to sanctify the decay of enlightenment thought; and mostly they are targeting the gradually developed and institutionalized socio-cultural networks and the democratic urban realm.

Although, the concept of techno-medieval is assigned to be an initial reflex for an alternative reading of this radical turn about the crisis of contemporary life, the consistency of both the concept and the concerns for delineating the *zeitgeist* affirmed its legacy some years later, in the spark of the global pandemic within the third decade of the twenty-first century. Similar to the emergence of the plague during the medieval century, with the emergence of COVID-19 in 2020, the technologically enhanced modern society of the ultimate century remarked itself as the most vulnerable being of the universe. Since we confront with a novel crisis in global scale and the measures of lockdown—getting heavier day by day—we are forced to remain restricted from the contemporary life and the urban realm. Even though the most-developed physical infrastructures and networks of the contemporary metropolitan realm became functionless at the end, while paving the way for a drained life within the cities, and the deserted metropolitan condition and the involuntary decay of related networks. In this respect, like the urban gateways of the medieval age overarching the centuries and reaching today, our smart gadgets become the contemporary technological-gateways, regulating the flow of pandemic society and the metropolitan life. So, as our corporal presences remain imprisoned in the micro-milieus or set-ups, while our techno-gateways became both the mediums of restriction, surveillance and control, as well as the transmitters of the images of this unknown and invisible oppression of the disease.

So, departing from struggles and crises of the contemporary metropolitan realm, either being effected via violence and terror of a ghostly medieval past and its revivalist apparitions, or through the surrendering external factors of diseases and many others, this chapter aims to unveil the continuous rise and fall of the urban-network utopias from the perspective of architecture and urbanism, via examining its last hundred years in accordance with the advent of modernist agenda

Thus, by validating its hypothesis about the "evident decay of modern condition and the ideologies of enlightenment or progressive ideologies, contrary to the technological advancements", this chapter further discusses both the discursive presence and critical role of architecture and urbanism in framing the modern metropolis and the network society, as well as the critical dialectic between the public sphere and the private realm.

While doing so, the exploration initiates via depicting the emergence of urban condition and its historical linkages, then focusing briefly on the theoretical discourse about metropolitan condition and the modern individual from the 1900s. Consecutively, mapping the continuities and discontinuities in the formation of contemporary urban condition, and the related networks, notion of public sphere, effecting socio-political conjectures are the goals of this chapter in approving its hypothesis on the continuous movements of rise and fall during the history and the focused period. By adopting critical and descriptive stances as its methodology, this work focuses on diverse variables affecting the metropolitan realm such as ways and means of production and technology, social transformation and changing life styles, which are either fuelling or sedating the advancements, struggles and crisis.

Accordingly, this chapter is developed in five consequent parts; the following part provides a brief historical framing and linkages about the modern condition of urban realm. The third part discusses the rise of utopias and the fall of the late-metropolitan condition, while in the fourth part dystopias, oppression and control are undermined. Finally, the critique of the contemporary condition is reflected by implying the end of an era—the modernist enlightenment—and its disappearing affects through focusing on the ethical and mental transformation of the individual and the society in relation to dark precursor of progress that operates through a subliminal level of suppression via deploying urban networks and technology. A significant emphasis is also held on the recent segregation of public and private realms in accordance with decaying networks, platforms of communication and affordance.

The Metropolis at a Glimpse: The Advent of the Big City and the Modern Individual, Dialectics of Public and Private Realm

Probably, beginning with the very earliest phases of the history, and reaching through the most sophisticated attainments of human existences, not only the emergence but also the gradual development of built environment and the rise of the cities have been closely shaped within the consequences of multiple interactions between power, production, technology, economy and social structures. Yet, technological advancements significantly impact the ways and means of production might be addressable as the leading factors in the development of the urban realm. So, the gradual shift from early agricultural societies, to merchant communities of medieval era, then from low-tech mechanical societies to electronic, and then to the digital, even to post-digital societies of today, has significantly shaped and changed both our perception and our practices of place making, as well as the strength and density of the networks and the nodes in between the urban sites.

Doubtlessly, throughout the history of human settlements, the phenomenon of urbanization or the advent of cities in a merely structured and systematic way with spatial-programme divisions might be clearly pinned back to initial settlements of the Neolithic Revolution.[3] The ruins of *Çatalhöyük* (7th Millennium BCE) in south Anatolia, and its precedent Göbeklitepe (10th Millennium BCE) in South Eastern Anatolia emerged as the very initial examples of the built environment in civilization history,[4] which were shaped in accordance with the initial

[3] V. G. Childe, "The Urban Revolution", in *Town Planning Review*, Vol. 21, No. 1, Liverpool University Press, 1950, https://www.jstor.org/stable/i40003353, Retrieved: 10 November 2020; V. G. Childe, Man Makes Himself, London, Watts & Co., 1936; M. E. Smith "V. Gordon Childe and the Urban Revolution: A Historical Perspective on a Revolution in Urban Studies", in *Town Planning Review*, Vol. 80, No. 1, Liverpool University Press, 2009, https://patterns.architexturez.net/system/files/MES-09-Childe-TPR.pdf, Retrieved: 10 November 2020.

[4] Please see, William L Langer, ed. *An Encyclopedia of World History* (5th ed.), Houghton Mifflin Company, Boston, MA, 1972; Andrew Curry, "Gobekli Tepe: The World's First Temple?" Smithsonian Magazine, 2008, https://www.smithsonianmag.com/history/gobekli-tepe-the-worlds-first-temple-83613665/ Retrieved: 7 November 2020; Oliver Dietrich; Notroff, Jens, "A Sanctuary, or So Fair a House? In Defense of an Archaeology of Cult at Pre-Pottery Neolithic

technological turn based on agriculture. The prevailing invention of the urban condition, remarkably the walled-city structure, was adopted as an archetype, visibly being pursued even in the ancient Greek city-states of *polises*.[5] This unitary emergence of the city as a habitable realm was further transformed into more complex pattern or urban networks in Roman period. The sovereignty and the military concerns were the effective factors of the systematic connection between the Roman cities, albeit the notable presence of trade routes—vastly the silk road—was the essential fact in the coherence of the urban-network scheme.[6] Yet, either on the rise or on fall of the socio-political structures of ancient Greek cities to Roman cities, then to Medieval city-states, and even approaching to the industrial turn, the socio-economical pattern of trade kept its importance in regulating the urban networks.[7] Like Tafuri discusses in his seminal work *Architecture and Utopia*,[8] the growing capital and network relations of late-Medieval merchant communities mostly in Italian peninsula and around the Mediterranean basin not only defined novel socio-political relations and classes but also transformed the existing urban patterns. While defining this moderate wealth accumulation and its transformative capacities under an umbrella term "early-capitalist society"—prior to the industrial turn—accordingly Kwinter also examined the rise of the urban sphere of the late-Medieval European cities.[9] He highlighted the evolving spatial features such as market places and clock towers as the elements of this pre-capitalist formations, so that the notion of time becomes an

---

Gobekli Tepe". In Laneri, Nicola (ed.). *Defining the Sacred: Approaches to the Archaeology of Religion in the Near East*. Oxbow Books, 2015.

[5] N. Fields, Ancient Greek Fortifications 500–300 BC (Fortress), Osprey Publishing, 2006; F. Yegül, D. Favro, *Roman Architecture and Urbanism: From the Origins to Late Antiquity*, Cambridge University Press, 2019.

[6] C. River, *The Incense Trade Route: The History of the Rise and Fall of an Ancient Global Economy*, Charles River Editor, 2019; R. McLaughlin, *The Roman Empire and the Silk Routes: The Ancient World Economy and the Empires of Parthia, Central Asia and Han China*, Pen and Sword History, 2020.

[7] I. Wallerstein, *The Modern World-System II: Mercantilism and the Consolidation of the European World-Economy, 1600–1750*, University of California Press, 2011.

[8] M. Tafuri, *Architecture and Utopia: Design and Capitalist Development*, MIT Press, Boston, MA, 1979.

[9] S. Kwinter, *Architectures of Time: Toward a Theory of the Event in Modernist Culture*, MIT Press, Cambridge, MA, 2002.

emerging concept of the urban network in medieval cities,[10] in addition to the former relations of politics, economy and production. For instance, late-Medieval cities like Sienna, San Gimignano (Fig. 11.2) or Luca were noteworthy milieus, where this socio-economic transformation was visible through the urban space, with the rising towers attached to the houses of wealthy merchant families.[11] These cities were also connected with each other in economic and cultural terms, while forming an inter-city network of relations[12] through which the knowledge for enlightenment was articulated. In fact, Medieval era was a significant duration, a period

Fig. 11.2 Historic Centre of San Gimignano Italy, Ko Hon Chiu Vincent, 2012. ([URL]: whc.unesco.org/en/documents/137468 UNESO World Heritage Center, Retrieved: 20th, 2021)

[10] Ibid.

[11] "Historic Centre of San Gimignano", in UN World Heritage Convention, https://whc.unesco.org/en/list/550 Retrieved: 30 October 2020.

[12] Though hundreds year later in his autobiography, Cellini still depicted the continuous tradition of economic and cultural interconnection between the cities, as well as the emergence of this cultural-network for the skilled artisans to attain commissions from these merchant families like Medici's. For further reading about the interval please see, B. Cellini, *The Autobiography of*

of transition that encompasses both the regressive and the progressive ideologies and occurrences in itself. Although the regressive ideology of dogmatic religious structures and the feudal governing mechanisms were effective mechanisms of decay and failure of the Antiquity and oppression of the society during the middle age; but like mentioned above the late-medieval era was the scène of change, a threshold introducing the rising values of early-modern urban condition, preliminary development of contemporary urban infrastructures, and the public sphere.

On the other hand, comparing to its historical precedents, evidently, the emergence of contemporary metropolis appears as relatively a new phenomenon within the urban history, mostly associated with the industrial turn and its massive and irreversible consequences in the ways and means of production and consumption chain.[13] Since the emergence of machinery empowerment of the humanity scaled-up the production and output capacities, this also densified the cities in terms of population, land use, development of advanced infrastructures, despite eviscerating the rural.[14] This significant technological and economic shift also transformed the socio-cultural infrastructures, individuals, and their perceptions, moral values and lifestyles, which all shaped the metropolitan condition at the time.

Accordingly, at the beginning of the twentieth century, by referring mostly to the transformation in and around Berlin, Simmel framed this ongoing transition in resonance with its multiple aspects. In his pioneering work *Metropolis and Mental Life*,[15] Simmel announced the reality of the emerging phenomenon or the metropolitan condition, which is totally different from both the former settlement typologies of rural environment. Simmel further introduced the rise of the modern

Benvenuto Cellini (org. 1562), Penguin Books, 2010.

[13] D.R. Meyer, "The Rise of the Industrial Metropolis: The Myth and the Reality" in *Social Forces*, Vol. 68, No. 3 1990, pp. 731–752. https://www.jstor.org/stable/2579351 Retrieved: 30 October 2020.

[14] As it is acknowledged in many scientific reports recently, the drastic reality of this longer transition asserts that the majority of the world's population is now living in the cities. See, "2017 Revision of UN World Urbanization Prospects", UN Department of Economic and Social Affairs Population Dynamics, https://population.un.org/wup/ Retrieved: 30 October 2020.

[15] Originally published in 1903, G. Simmel "Metropolis and Mental Life", in *Simmel on Culture: Selected Writings*, D.P. Frisby and M. Featherstone (eds.) SAGE Publications Ltd, 1008.

metropolitan individual, while depicting its struggles and crises within this novel set-up. This seminal work examined the macro and micro relations of the industrial turn, via dismantling the interconnection between the industrialized modern metropolis, public sphere and the individual psyche in the modern society. Simmel further theorized their dynamic and interchangeable capacities. He emphasized that although the contemporary metropolis of the 1900s was shaped under the decisive factors of economic, political, technological social networks and accommodates them, but the metropolis itself is exactly what these relations are.[16] Thus, the contemporary metropolitan condition emerges through and under these active vectors. So, departing from these unstatic, but rather dynamic and interchangeable characteristics of the metropolis, Simmel also discussed the novel condition of the metropolitan individual in comparison to the archaic rural individual. This assessment depicted the shifting interconnections of the closed and open societies in the vital presence of communication and metropolitan socio-cultural networks.[17]

In this respect, during the gradual integration and transformation of the individual into a modern metropolitan citizen, the relief from the former socio-cultural and economic ties emerged as the initial step in affirming the individual independence.[18] However, this emancipation soon declined with the failure of the modern individual being trapped with the metropolitan psyche, which in fact transforms the individual and the public sphere through loosening bonds and interactions between the individuals. Like Simmel argued that this failure would pave the way to estrangement of the individuals to the milieu—the environment and the society—that they exist.[19] So, they recessed into their protective clusters—the *interiors*—from the "fluctuations and discontinuities of the external milieu threaten [them]".[20] In this way, for the modern individual, the metropolitan condition and its related networks acted as a liminal space existing between the phases of recognition and estrangements.

[16] G. Simmel, Metropolis and Mental Life, 1903.

[17] Ibid.

[18] Ibid.

[19] Ibid.

[20] Ibid., p. 12.

Moreover, Simmel also introduced the notions of both perception and affect in individuals' withdrawals and integrations to the modern city life, as well as articulating crisis of transformation. In other words, perception and affect became mediatory notions that were regulating the integration of the individual in sequential steps. On the initial phase of "implication" the city's affective stimuli were articulated unevenly in the human body, which would cause the processes of recording and inward bending or folding of these metropolitan exposures. Then on the continuing step of "complication", the stimuli created by the differences between these recorded effects and the individual's enhanced interaction capacity allowed the individual to engage and resonate with possible interaction potentials that appear around it as well as enabling the individual to develop new activities. As stated by Simmel this was also the process of twisting creation of novel networks or bonds that the stimuli crated when they were being filtered at that moment. And finally on the last step "explication", those stimuli that were filtered at that moment were scattered back to the metropolis as transformed-new vectors, which were the outward bending process that the affects return to the flow of life again.[21]

In short, despite the physical infrastructures and the modern built environment, of complex relations of production and economy, Simmel significantly addressed the importance of mental networks and their transformative capacities in the constitution of metropolitan realm and the modern individual inhabiting it. Without a doubt, these mental networks were effective through the presence of communicative networks of the edificatory metropolitan condition in the progressive mood of the modernist thought. The other significant detail that Simmel portrayed was the emergence of micro-milieus. These were environments that the modern individual constructed to survive from multi-fold processes of integration or disintegration. Evidently this micro-space that Simmel introduces was obviously beyond the connotations of a conventional domestic realm or interiority, but like mentioned above, merely multiplied liminal spaces of the subjectivation where the modern individual realizes itself, while interacting through diverse channels of communicative networks within the city. Similar to Simmel, Borges also revealed this

[21] Ibid.

micro-milieus of existences, and the associating cognitive processes in relation to the interaction between the public and private realm in the contemporary metropolitan life. Approaching merely from a confined perspective, Borges introduced his notion of labyrinth as the invisible penitentiary of the modern individual, who is supposed to be liberated and mobile, but being stuck in the constraints of this invisible micro-structures and being atomized in the public realm.[22] Thus, Borges discusses the detainee condition of the modern individual and its psyche resonating not only with the contemporary socio-political life and the modern city, but also with power relations and sovereignty.[23] Borges argued that the modern individual carries the labyrinth with itself wherever it moves,[24] without requiring any physical reference of containment. In other words, this is an abstract spatiality of imprisonment, a mental state that transfigures in relation to diverse situations and the interaction of the individual with the infrastructures of the urban realm.

However, the micro-milieus of the existence either being confined through individual or force major conditions, or being extrapolated and enriched with multiple processes, there emerged a significant fluidity of connections between the two realms of private and the public. Like the emergence of micro-milieus of existence in relation to metropolitan transformation, the emergence of modern public space was also shaped mutually with the private realm.

The historical linkages of contemporary public sphere in terms of its intellectual potentials and its associating mechanisms might be traceable back to the era of Enlightenment. Thus, despite its earlier connotations with production and exchange emerged in the late-Medieval era, cognitive, communicative and intellectual development of public sphere—within the urban realm—mostly emerged with the progressive ideology

[22] E. Redekop, "Labyrinths in Time and Space", in *Mosaic: An Interdisciplinary Critical Journal*, Vol. 13, No. 3/4, Other Worlds: Fantasy and Science Fiction Since 1939 (Spring/Summer 1980), University of Manitoba, 1980, pp. 95–113, https://www.jstor.org/stable/24780265, Retrieved in January 16, 2021.

[23] J. Hahn, "Labyrinth, the Shape of the Modern Mind: Kafka, Auster, Borges, Wellesley College, Honors" in *Comparative Literature*, Wellesley College Digital Scholarship and Archive, 2013. pp. 55–77 http://repository.wellesley.edu/cgi/viewcontent.cgi?article=1143&context=thesiscollection. Retrieved in July 1, 2020.

[24] Ibid.

and social transformation beginning with the Enlightenment. By introducing the notion of *Öffentlichkeit*[25] Habermas emphasized the rise of the democratic public sphere in Europe, during the era of Enlightenment. He further highlighted the growth of intellectual capital in the modern urban realm, so that the logic and "rational discussion takes place between citizens on matter of general interest", whereas the public opinion rises from these formal or informal debates, and their influential capacity in organizing the society.[26] This intellectual transformation of the individual on behalf of progressive stance gave way to the formation of novel milieus and interfaces of gathering and meeting in the public sphere[27] as well as the articulation of social structures in the city. Thus, like Habermas referred the appearance of new actors of the modern society—such as the educated bourgeois class of doctors, lawyers and scholars—in a way expanded the novel public spaces into liminal spaces, the temporal platforms of intellectual exchange and communication.[28] These temporal milieus of exchange were being hosted by the physical interfaces of novel typologies of modern city like "salons, cafes and clubs, where members of different classes met to engage in debate, verbal sparring, and display of rhetorical sophistication".[29] In addition, increasing importance of freedom of expression further challenged the birth of numerous impersonal actors of communication network, such as newspapers, books, periodicals and so on,[30] contributing to the development of the public sphere.

In negation to Habermas, philosophers like Negt and Kluge emphasized the essence of the *counter*—the *other*—through expanding the public sphere beyond the affordance of the capitalist and edificatory cycles of

[25] J. Habermas, *The Structural Transformation of the Public Sphere: An Inquiry into a Category of Bourgeois Society*, MIT Press, Cambridge MA, 1989.

[26] T. Avermaete, "Architecture as Res Publica: The Public Realm as Frame, Substance and Goal for Design", in *Delft Lecture Series Architectural Design* Susanne Komossa, Roberto Cavallo (eds.), TU Delft Press, Delft, 2014, pp. 28–45.

[27] Ibid.

[28] *The Structural Transformation of the Public Sphere: An Inquiry into a Category of Bourgeois Society*, 1989.

[29] T. Avermaete, "Architecture as Res Publica: The Public Realm as Frame, Substance and Goal for Design", 2014.

[30] Ibid.

the social structures and classes of the modern city.[31] Negt and Kluge elaborated the dynamic structures of the public life, and implied the constant emergence of new forms and means of public sphere that would be heterogeneous and not to be assigned into privileged classes or groups.[32] While resonating with Negt and Kluge, and criticizing Habermas' interpretations of the public sphere, Fraser developed a more comprehensive definition of the public sphere in the metropolitan realm, by including the invisible actors in the formation of the broader social infrastructure.[33] Almost resonating with Foucauldian perspectives, Fraser's actors were the ones who were discarded and segregated from the sterile urban life due to their gender, race or status.[34]

Moreover, resonating with these propositions about the definition of the public sphere in the contemporary realm, Arendt re-introduced the political dimension of the public sphere one more time. Referring to the self-realization process of the individual, Arendt defined the public realm as the *spaces of the appearances*,[35] where the individual claims its existential presence through its multiple interactions with the social infrastructures of the city. Unlike the limited labour and space interaction of the late-capitalist society, Arendt also acknowledged the public realm as a milieu that people act, rather than work.[36] According to Arendt, only through these intermingling events of action and speech people can create a spatiality in-between each other. According to Arendt, this liminal space of existence becomes the place like she stated "I appear to the other and the other appear to me".[37] What is noticeable in Arendt's discussions was the emphasis that she drew up on, not only the interaction between the

[31] O. Negt and A. Kluge, *Public Sphere and Experience: Toward an Analysis of the Bourgeois and Proletarian Public Sphere*, University of Minnesota Press, Minneapolis, 1993.

[32] Ibid.

[33] N. Fraser, "Rethinking the Public Sphere: A Contribution to the Critique of Actually Existing Democracy", in *Habermas and the Public Sphere*, Craig Calhoun (ed.), MIT Press, Cambridge: MA, 1992.

[34] Ibid.

[35] H. Arendt, *Human Condition*, University of Chicago Press, Chicago, 1958.

[36] Ibid.

[37] Ibid., p. 198.

individuals in the public sphere, but also the broader integration of all "other living and inanimate things" that come together to make this wider sphere as the public realm. Arendt also undermined the importance of visibility and accessibility of everything and each one as the basics of the democracy in the public realm, and fostering the commonality.[38]

In fact, the emphasis on communality addresses a slight, but a very crucial distinction between the notions of *public space* and the *common space*, whereas the latter enfolds the democratic essence that plays the active role in self-realization of the individual being provided an inclusion of this collectivity that it emerges. Since this avoids the individual being "remained suspended [with]in its own individuality",[39] it further paves the way in building collective individual processes. The notion of common spaces in fact appears as an activist redefinition in the current condition of the post-metropolitan condition, in order to democratize the public and private polarization, on behalf of developing "an open and inclusive meshwork, where the multitude freely shares, utilizes and sustains a pool of commonwealth based on forms of participatory self-organizations, … [through]constantly renegotiating, redefining and reproducing their social commonality",[40] through embracing the versatile differences. Like we have been witnessing through diverse performative actions beginning with occupy New York or Arab Spring and its consequences, in terms of claiming the commons against the sovereign powers of late-capitalism, socio-political and autocracy, the notion of common space clearly designates the final stage of the intertwined relations of metropolitan transformation, its related infrastructures, the crisis and limits of the contemporary urban condition today.

[38] T. Avermaete, "Architecture as Res Publica: The Public Realm as Frame, Substance and Goal for Design", 2014.

[39] Ibid.

[40] G. Kodalak, "A Monstrous Alliance: Open Architecture and Common Space" in *Footprint Journal*, Vol: 9 No: 1, Commoning as Differentiated Publicness, Spring 2015, pp. 69–89. https://doi.org/10.7480/footprint.9.1.900 Retrieved: January 16, 2021.

The Rise of Utopias: Projecting the Spatio-politics of the Modernist Network Society

As mentioned earlier, the evident impact of technological, political, social and economic changes comprehensively altered the micro-geographies of individual in terms of cognitive and habitual aspects, the public and private dialectic in mezzo-scales of the urban realm, finally the macro-geographies of districts and metropolitan networks. This zeitgeist of transformation—initially the industrial turn, then the other developments—proposed milestone projects focusing on presents and futures of the cities, while offering better conditions. These projects emerged as the dispositive of almost utopic envisions, that prioritized unprecedented necessities in designing the modern city.

On the other hand, in his book entitled *Architecture and Utopia* (1979)[41]—a Marxist critic on architecture and urbanization—Tafuri criticized the agency of architecture and its centuries-long role in masking the "underlying contradictions of capitalist society".[42] So, Tafuri emphasized the essence of utopias as "concerned with the question of the rational organization of society—overcoming the contradictions inherent in the capitalist society".[43] According to Tafuri, to maintain the justice, utopias tried to organize the society in a rational manner with the accepted logic of development.[44]

So, these utopic projects of the twentieth century mostly claimed a total breakthrough from the pre-modernist tradition, whilst proposing a clear start up in forming the visionary urban condition. In this way, utopic projects emerged as precursors of progress, blinking in the crisis of the contemporary life or during the shifting conditions of the urban realm in the last century. In other words, they were the significant milestones, in defining the route for shaping the spatial composition and the infrastructures of our contemporary network society. In this respect,

[41] M. Tafuri, *Architecture and Utopia: Design and Capitalist Development*, MIT Press, Cambridge, MA, 1979.

[42] F. Torisson, *Utopology a Re-Interrogation of The Utopian In Architecture*, Doctoral Dissertation in Architecture, Lund University, 2017.

[43] Ibid.

[44] Ibid.

three visionary projects from critical thresholds of the twentieth century were mapped as the cases of discussion, in terms of reflecting how their propositions were closely tight with the existing inquiries of urban space, sociology and existential theories, struggles and opportunities of capitalism and technology, production and theories of economy. Accordingly, early 1900s and crisis of industrialization in cities, post-war recovery and the re-making of the metropolis, cybernetic turn and the affirmation of network city and the control society were the three evident intervals of the evolutionary processes in shaping the contemporary network society.

The evident start of building the cities in accordance with the industrial turn might be addressable with the visionary projects of both Howard and Garnier: initially the book *To-Morrow: A Peaceful Path to Real Reform* (1898) (Fig. 11.3) and then the proposal of "Garden City" by Howard (1902), as well as "The Industrial City" (Fig. 11.4) project by Tony Garnier (1901).[45]

Ignoring the problems of historically articulated existing cities, and to start building the cities of industrial turn, both Garnier and Howard suggested clear beginnings by occupying novel territories independent of the existing settlements. As both Howard and Garnier introduced novel aesthetical relations of space and form, they also prioritized the infrastructural networks as the core components of the system, connecting the diverse and multiple relations of production, governing, education, housing and leisure programmes and their dedicated zones on the metropolitan in the urban footprint and hierarchy. Basically the proposal encompassed what was needed in a city through providing an advance network of the mentioned infrastructures.

Doubtlessly, the interventions of both Garnier and Howard positioned their basis on the optimization and production efficiency of the capitalism, while proposing a moderate automation of the urban condition through the integrated networks of diverse infrastructures. Thus, the definition of space for workers' houses and the broader urban network of the

[45] E. Howard, *Garden Cities of To-morrow*, London: S. Sonnenschein & Co, 1902, http://library.lol/main/59EB59F5FD9AC4B783C0C2EBE6BB7E76, Retrieved: January 14, 2021. D. Wiebenson, "Tony Garnier: the cité industrielle (Planning and cities)", *Studio Vista* (January 1, 1969). D. Wiebenson "Utopian Aspects of Tony Garnier's Cité Industrielle", *Journal of the Society of Architectural Historians* Vol. 19, No. 1 (Mar., 1960), University of California Press, https://www.jstor.org/stable/987962, Retrieved: January 14, 2021.

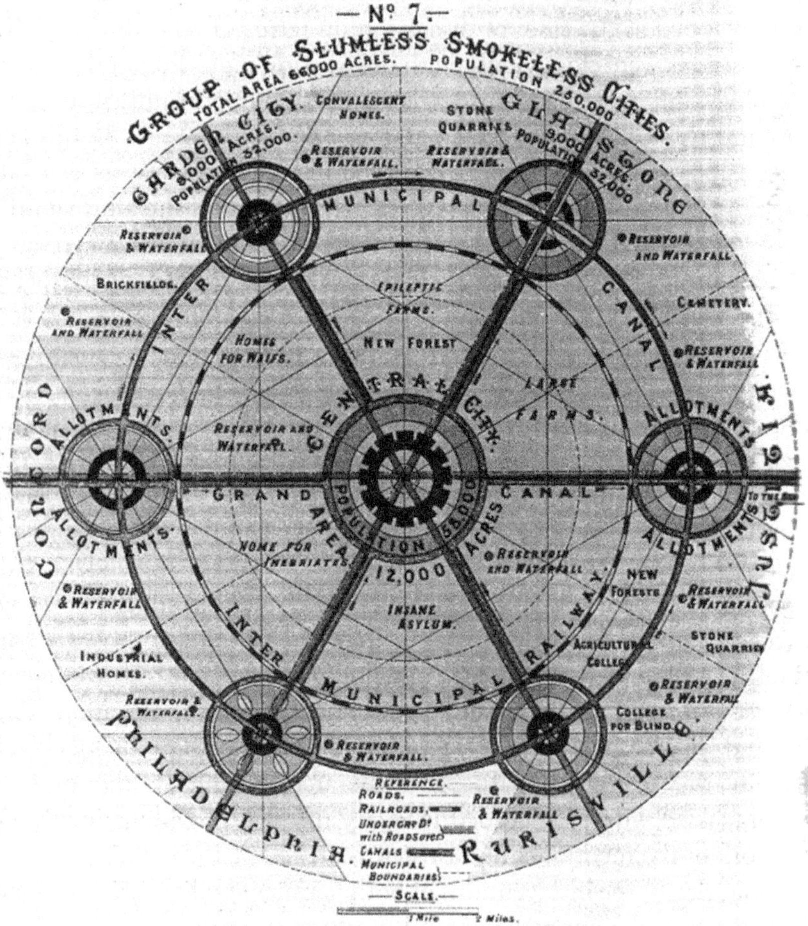

Fig. 11.3 Garden City Scheme, Ebenezer Howard, 1898. ([URL]: https://cargocollective.com/artifact/Garden-City-the-Green-Metropolis, originally published in Howard E. Tomorrow: A Peaceful Path to Social Reform, 1898. Retrieved: 20th, 2021)

modernized city were almost scheduled to increase the productivity. In this respect, workers' accommodation was placed in the vicinity of the production centers by the periphery of the city, enabling efficiency and affordance in their transportation on behalf of reducing the time and

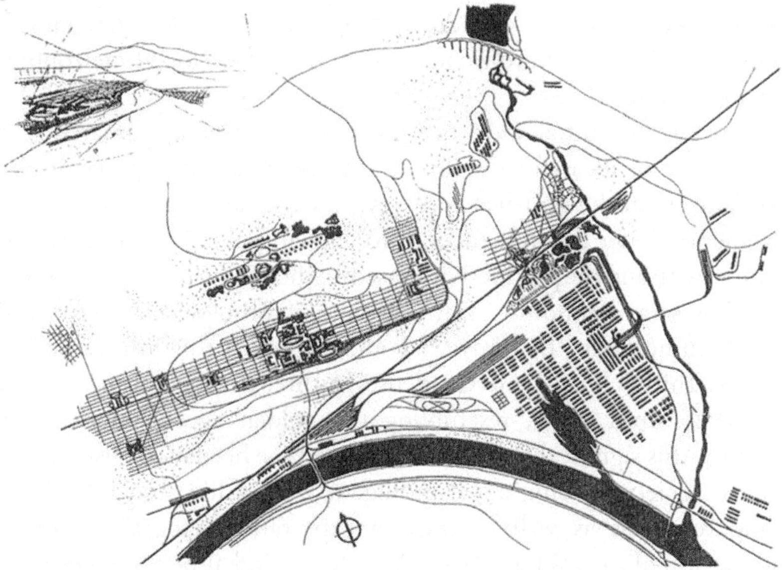

Fig. 11.4 Industrial City Plan, Tony Garnier, 1901. ([URL]: https://www.penccil.com/gallery.php?p=490504414159, originally published in Garnier T., *Une Cité Industrielle*, 1917. Retrieved: 20th, 2021)

resource consumption in commuting and increasing the travel safety.[46] Thus, from a relatively socialist perspective, the schema of industrial cities utopia strongly sought the further accessibility of its citizens directly to the essential public services such as healthcare, nurseries, education infrastructures, as well as the related leisure infrastructures of individual and collective well-being of the inhabitants such as sport centres, pools, beaches, entertainment bars, night clubs, restaurants, movie theatres.[47] In this way, the proposed utopias of the industrial turn imitated a dispositive of an ideal city, being designed to fulfil all the demands of its inhabitants. Although, this refabricated image of oasis might be asserted in relation to basic instinctive features of social (belonging) and teleological (heavenly

[46] Ibid.

[47] Ibid.

existence)[48] concerns, but apparently both projects—of Industrial City and the Garden Cities—seriously proposed series of interconnected physical and infrastructure web for projecting a developed metropolitan network system.

Following the initial wave of industrial city utopias, apparently the second wave of ideas for developing the modern city reposed itself in accordance with the post-war recovery processes and its rising economic movement. Under the influences of Keynesian policies to boost the economic growth and the productivity based on increasing the consumption,[49] the throw-away economy or the notion of obsolescence became the essential indicator of this wave.[50] Foursatié defied this interval as "The Glorious Thirty" (Les Trente Glorieuses) a three decade of time spanning from 1945 until 1975, resembling the golden age of modernism.[51] Evidently, this significant interval designated the heyday of post-war consumerist society desiring a self-indulgent way of life, and full of ever-ending resources as well as feasting expectations, together with the mobilization of the middle classes. Yet, this period also marked the duration of recovering post-war societies through the construction of diverse infrastructures of modernization and social transformation. The trans-Atlantic urbanization policies mostly affected the preferring countries in

[48] M. Pimlott, "Fiction and Significance in the Public Interior: Culture & (Its) Imagination: Spaces and Methods of Place Making", in *Delft Lecture Series Architectural Design*, Susanne Komossa, Roberto Cavallo (eds.), TU Delft Press. Delft, 2014, pp. 28–45.

[49] A. J. de Regil "Keynesian Economics and The Welfare State", in *Neo-Capitalist Assault*, the Jus Semper Alliance, N:4, 2001, pp. 1–19, https://www.jussemper.org/Resources/Economic%20Data/The%20Neo-Capitalist%20Assault/Resources/KeynesianEconomics.pdf, Retrieved: January 16, 2021.

[50] N. Whiteley, "Toward a Throw-Away Culture. Consumerism, 'Style Obsolescence' and Cultural Theory in the 1950s and 1960s", Oxford Art Journal, 1987, https://ccnymadmen.files.wordpress.com/2016/07/toward-athrow-away-culture.pdf Retrieved: January 18, 2021.

[51] For the original work of Foursatié please see *Les Trente Glorieuses, ou la Révolution Invisible de 1946 à 1975*, Paris, Fayard, 1979. For English interpretations of the work, please see N. Crafts and G. Toniolo "Les Trente Glorieuses: From the Marshall Plan to the Oil Crisis", Oxford Handbooks Online, 2012, https://doi.org/10.1093/oxfordhb/9780199560981.013.0018. Retrieved: July 15 2020; E. Hobsbawm, *Age of Extreme: The Short Twentieth Century 1919–1991*, Abacus Book, London, 1995.

Europe[52] that also became the beneficiaries of the global aid plans,[53] while allowing the auto-mobilization of their cities.[54] In this way, the emergence of motorway infrastructures as the essential metropolitan networks[55] and novel planning decisions based on the linear growth of the cites for the endless development and consumption became the essential acts in the planning of post-war metropolitan realm.[56]

In accordance with immense tendencies of obsolescence, global capitalism's further aspects towards increasing the mobility of the middle class and the car ownership, and the emergence of motorway infrastructures as the veils of the modern city, totally transformed the modern individual's perception of the city around the 1960s. In this respect the notions of autonomy, mobility, accessibility, interchangeability and flexibility appeared as the key concepts of the time penetrating to the evolving production, service, communication and leisure networks of the city.[57] Besides, in resonance with the democratic distribution of the resources the rising impact of welfare state lit the second escalation of utopic envisioning.

For instance, the interrogative projects like "Walking City" (1964), "Plugin City" (1964) or "Instant City" (1968–70)[58] (Fig. 11.5) by an architecture collective *Archigram*[59] were significant cases that criticized

[52] For the cases of Sweden and Turkey please see, H. Mattsson, "Were the Motorways Meet: Architecture and Corporatism in Sweden 1968", in *Architecture and the Welfare State*, M. Swenarton, T. Avermaete, D. van der Heuvel (eds.), Routledge, New York, 2015, pp. 155–175; I.Y. Akpınar, "The Making of a Modern Pay-I That in Istanbul: Menderes Executions After Prost's Plan", in *The Imperial Capital to the Republican Modern City: Henri Prost's Planning of Istanbul (1936–1951)*, C. Bilsel, P. Pinon (eds.), İstanbul Research Institute, Istanbul, 2010, pp. 167–199.

[53] L. I. Bland, "Marshall and the Plan", The George C. Marshall Foundation, https://www.marshallfoundation.org/marshall/the-marshall-plan/history-marshall-plan/marshall-and-the-plan-by--larry-i-bland/, Retrieved: 18 January 2021

[54] A. Smithson and P. Smithson, "Mobility: Road System", in *Architectural Design*, October 1958.

[55] B. Adalet, *Hotels and Highways: The Construction of Modernization Theory in Cold War Turkey*, Stanford University Press, 2018.

[56] A. Smithson and P. Smithson (eds.) *Team 10*, Studio Vista, London, 1968; M. Risselda and D. van den Heuvel (eds.) *Team 10*, NAİ Publishers, Rotterdam, 2005.

[57] P. Cook (ed.), Archigram, London, 1972.

[58] S. Sadler, *Archigram: Architecture Without Architecture*, MIT Press, Cambridge MA, 2005.

[59] Archigram is an architecture collective, "preeminent architectural avant-garde of its time" almost a decade in 1960s; focusing on architecture, theory, criticism, design and urbanism and founded in 1961. The group consists of six members including Warren Chalk, Peter Cook, Dennis Crompton,

Fig. 11.5 Instant City, Archigram, 1969. ([URL]:http://archigram.westminster.ac.uk/project.php?id=119. Retrieved: 20th, 2021)

these existing modalities of the time, while offering a novel set of spatial relations[60] in transforming the future of thc cities. In this way, the collective proposed edificatory urban practices, via targeting the entertainment and leisure facilities as the immanent and subliminal networks of information and education for the development of modern individual.[61]

David Greene, Ron Herron and Michael Webb. S. Sadler, *Archigram: Architecture Without Architecture*, 2005.

[60] Although the architecture that the group presented has been also categorized as the neo-avantgarde architecture of the 1960s the projects are defined as utopias and framed as visionary approaches fuelling the progressive impact of development. For neo-avantgarde reading of the projects please again see, S. Sadler, *Archigram: Architecture Without Architecture*, 2005.

[61] "Peter Cook (Archigram): Instant City, 1968–1970", FRAC Centre-Val de Loire https://www.frac-centre.fr/_en/art-and-architecture-collection/cook-peter/instant-city-317.html?authID=44&ensembleID=113 Retrieved: December 13, 2020.

While introducing "the idea of traveling metropolis",[62] Archigram projected a set of sequential temporal activities—that were embedded in the mobile and temporal bodies of the instant spatialities—instantly penetrating to the communities, where these travelling cities visit or connect. In this respect, the Instant City project, benefited from the contemporary achievements of the technological advancements, while proposing a mobile embodiment augmented with airships and hot air balloons. Apparently, the project also undermined the prospect possibilities of exoterrain development of a city, being truly ungrounded from the conventional urban realm. On the other hand, from a perspective of a disappearing architecture, Instant City prioritized construction of the image and the series of intangible atmospheres, rather than stiff and tangible embodiment of an architecture. Accordingly, the architectural embodiment was programmed almost intangibly, that would be thoroughly dissolvable in the local communal networks. Evidently, Instant Cities would almost act as access terminals in the broader metropolitan network chain. Deliberately, the feeling of thoroughly being inside and immersed to the experience of subliminal learning via experimental nature of entertainment and leisure were the essences of the invention, while altering the duality or developing novel definitions in the dialectics of public and private realm, as well as the relations of society and the individual. In this respect, on behalf of the intense knitting of my statements into each other, and bridging the discussion within the text, it is worth mentioning the consequent appearance of Arendt's reference of communality or the projective ideas of the Instant City, or Negt and Kluge's statements for accommodating the *other* in terms of uplifting every members of society were not random occurrences. Although there were no direct evidences nor statements about the group's inspiration or references to the theories of neither Arendt nor Negt and Kluge on public sphere, but obviously Archigram's interrogations in Instant City for a novel propositioning of the communality or publicness were evidently overlapping with the mentioned theories.

Pursuing these attainments, the computational turn and the advent of cybernetics around late 1960s address the third significant threshold in

---

[62] Ibid.

the utopic projects. In his seminal work *The Architectural Relevance of Cybernetics* (1969) Pask introduced the cybernetics into the realm of spatial practices "as a compilation of active systems".[63] So, through elaborating with the inquiries of ontological presence and utilizing programmability and flexibility of spatial, Pask advocated immateriality and transformability architecture from a reconfigurable perspective.[64] Pask also prioritized the autonomy and emancipation of this novel place making practices beyond their dependency on human. He further proposed a novel system of architecture that could "calculate, determine and predict" and optimize design processes.[65] Echoing towards today's AI discussions, and before the significant critique of control society, Pask's intervention of cybernetics in architecture appeared as an invention for machinery decision-making processes of incorporated multitudes on behalf of architects, for the sake of the user.[66] Thus, the architectural relevance of cybernetics proposes a competence on evaluation based on an open ended system, continuous learning from itself.[67]

Doubtlessly, the reflections of the cybernetic turn lasted for longer durations not only in architecture but also in diverse processes of urban studies. Despite the evident impact of cybernetics in the 1960s—in the context of second-wave utopists like Cerdic Price[68] and Archigram[69]—a

[63] G. Pask, "The Architectural Relevance of Cybernetics", in Computational Design Thinking, pp. 68–77. https://cmusyntheticecologies.files.wordpress.com/2014/01/pask_gordon_the-architectural-relevance-of-cybernetics.pdf Retrieved: January 10, 2021.

[64] Ibid.

[65] Ibid.

[66] Ibid.

[67] Ibid.

[68] Similar to the mentality of second-wave utopists, the inspiring work of Cedric Price Fun Palace (1964) also introduces the edificatory roles of the entertainment and leisure infrastructures in the broader articulation of metropolitan social networks. For more information on the project please see, S. Mathews, "The Fun Palace: Cedric Price's Experiment in Architecture and Technology", in *Technoetic Arts: A Journal of Speculative Research*, Vol.3 No.2, Intellect Ltd., 2005, https://doi.org/10.1386/tear.3.2.73/1.

[69] Although the direct dialogue and the exchange of ideas between Pask and Archigram members—like Peter Cook—were acknowledged, author is repositioning the cybernetics theory in relation to the third wave and in order to prioritize its after-impacts while enhancing the discussion. For the emergence of cybernetics in architecture curricula in Architectural Association-London, please see F. Torisson, "The Cybernetic Hypothesis & Architecture" in Histories of PostWar Architecture, No.1, 2017. https://doi.org/10.6092/issn.2611-0075/7208, Retrieved: January 18, 2021.

decade later its resonance on the succeeding utopic projects was also remarkable. For instance, the works of Takis Zenetos "Electronic Urbanism"[70] addressed in this respect as a novel wave of envisioning" (Fig. 11.6). Being a member of the International Cybernetic Association, Zenetos unveils the possibilities of a fascinating future of the city "detached from the ground as a cloud colony", utilizing the vertical sprawl and electronic communication devices in order to dwell in the atmosphere.[71] Evidently, through appreciating the immersive presence of telecasting and communication networks, the project hinted the omnipresence of life through these infrastructures and freed from the conventional materiality.

In short, the utopic envisions of the twentieth century were responses to the restrictions and impasses of the developing capitalism, while being benefited from the achievements or opportunities of its intermingled processes of technology, production and economy. They instrumentalized these intermingled processes to propose positive impacts to transform the

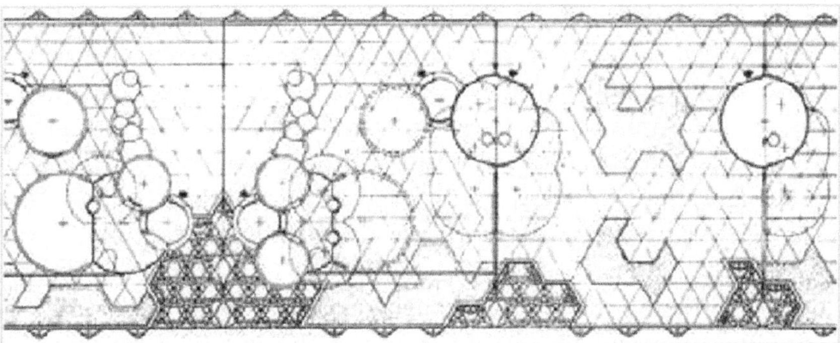

Fig. 11.6 Electronic Urbanism, Takis Ch. Zeneto, 1974. (L. Kallipoliti, "Cloud Colonies: Electronic Urbanism and Takes Zenetos' City of the Future in the 1960s", ACSA Annual Meeting Proceedings, 2014. https://www.acsa-arch.org/proceedings/Annual%20Meeting%20Proceedings/ACSA.AM.102/ACSA.AM.102.77.pdf, Retrieved: January 16, 2021. originally published in Zenetos, T. Ch., "Town Planning and Electronics" in *Architecture in Greece, Annual Review*, No.8, Athens: 1974)

---

[70] L. Kallipoliti, AA Files No.77, 2020, pp. 18–27. L. Kallipoliti, "Cloud Colonies: Electronic Urbanism and Takes Zenetos' City of the Future in the 1960s", ACSA Annual Meeting Proceedings, 2014.

[71] Ibid.

society, and to enhance the urban condition and the related networks in order to increase the opportunities for the citizens.

Rising Dystopias: Emergence of Oppression and Control and the Decay of Urban Realm and the Network Society

Apparently, the most appropriate definition of the term dystopia is Foucault's innovative definition of the notion in relation to politics of space and related regimes of oppression for the re-organizing of spatiality.

Probably, Hausmann's implementations for the reconfiguration of Paris into a modern metropolis addressed the significant encounter of the modern individual with the biopolitics and the utilization of urban space and especially the public sphere and the infrastructure networks as a tool to maintain the control society.

The impasses of the late nineteenth-century city life and the struggles of the sovereignty between the state and the resisters were the key consequences to operate the formerly structures—almost to the Medieval origins—of the city on behalf of regulating and "sanitizing" the urban realm. More specifically, Hausmann's two-fold intervention—both underground and above-ground (Figs. 11.7 and 11.8)—for the reconfiguration of the city was a direct intervention to maintain states' control over the city and the citizen, in order to avoid further resistive impulses of the "others". Striating the spaces through the networks of transportation, communication and the sewage infrastructure were the key steps in homogenizing the metropolitan realm. In other words, from a Deleuzian perspective of disjunctive synthesis, Hausmann's interventions operated in an institutional level with an authoritarian tone, in terms of defining molar structures of the city, in order to surveill and control the flow in various levels.

Referring to Foucault, Vidler stressed both the economic and spatial stratification of the society and the urban realm, immediately paved the way to the segregation in between the classes and ceased the togetherness

Fig. 11.7 Illustration of nineteenth-century Boulevard Richard-Lenoir, J. Gaildrau, 1860s. (Illustration of nineteenth-century Boulevard Richard-Lenoir, J. Gaildrau 1860s, Benevelo, L., *The Origins of Modern Town Planning*, MIT Press, Cambridge: MA, 1971. originally published in Alphand, A., *Les promenades de Paris: Histoire—Description Des Embellissements—Dépenses De Création Et D'entretien Des Bois De Boulogne Et De Vincknnes Champs—Elysees—Parcs—Squares—Boulevards Places Plantees Etude Sur L'art Des Jardins Et Arboretum* (Fig. 347, p. 239))

of the society in terms of social control.[72] So, Vidler argued that, this split between the social classes and their positioning through diverse milieus of existences in the city in fact stems from not only the relations of capital and intellectual specialties of the citizens,[73] but also from the power relations, which causes the fragmentation of the public sphere[74] and fuels the resistance.

Apart from the political factors, around the 1970s, the immense environmental catastrophes—the float in Florence, the emergence of the

[72] A. Vidler, "Spatial Violence", in *Assemblage*, No.20, 1993, 84–85.

[73] G. Simmel, Metropolis and Mental Life, 1903.

[74] A. Vidler, Spatial Violence, 1993.

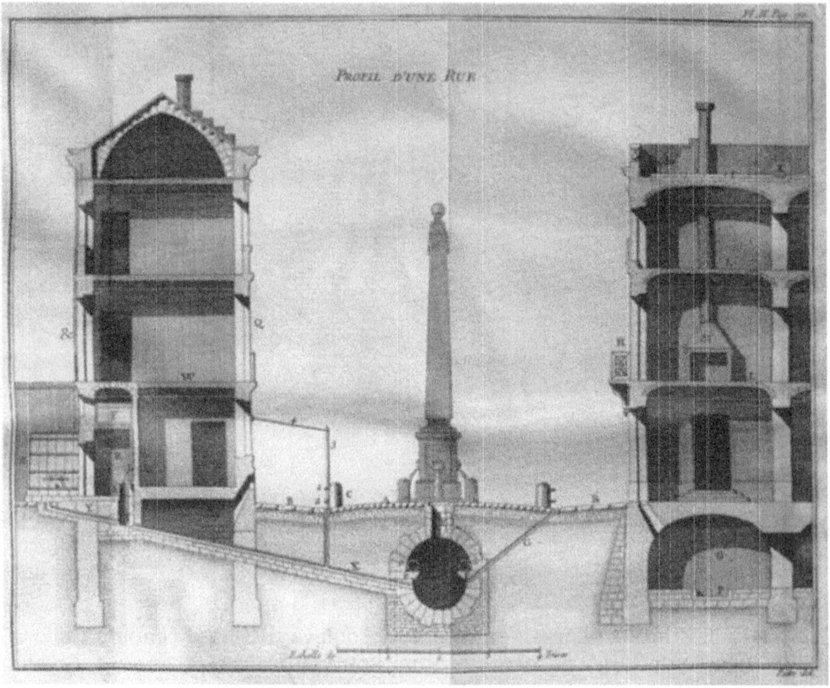

Fig. 11.8 "Profil d'une rue", Pierre Patte's street section Pierre Patte 1769. (Tallon, A. J., "The Portuguese Precedent for Pierre Patte's Street Section", in *Journal of the Society of Architectural Historians*, Vol. 63, No. 3, University of California Press, 2004, pp. 370–377, http://facultysites.vassar.edu/antallon/pdfs/tallon-patte.pdf Retrieved: January 16, 2021. Originally published in, Patte, P. *Mémoires, Sur Les Objects Les Plus Importans De L'Architecture*, Rozet Library, Paris, 1769 pl. 2, https://gallica.bnf.fr/ark:/12148/bpt6k5701519t/f4.item.texteImage, Retrieved: January 2021)

energy deficiencies—the fuel crisis 1970s and increased militarization—the Vietnam War and the consequences of the cold-war 1968 were other significant threshold, effecting the dystopic projections. Accordingly, drastic results of the Meadows' report *The Limits of Growth* (1968–71)[75] about the decaying future of the world resources also appeared as the catalyser of the mid-century dystopic turn. Significantly, the emergence

[75] D. H. Meadows, D. L. Meadows, J. Randers, and W. W. Behrens, *The Limits of Growth*, A Potomac Associates, 1968–71.

Fig. 11.9 Continuous Monument, Manifesto New York, Superstudio, 1969. ([URL]: https://www.frac-centre.fr/_en/index-authors/rub/rubprojects-317.html?authID=185&ensembleID=988&oeuvreID=13218, Retrieved: January 20, 2021)

of Italian radical architecture groups was decisive in the development of the dystopic envisions of this time. In this sense, through instrumentalizing powerful images and mostly handmade-collages, the alienating affects of Superstudios[76] were very interrogative. In their two seminal works "Continuous Monument" (1969) and "Camp" (1970) (Figs. 11.9 and 11.10), the groups examined the un-crystalized and horizontal structures of history writing via re-designability of the buildings and structures within the city.[77] In this way, Superstudio experimented the presence of a novel reality. Through their provocative works Superstudio claimed "The

[76] Superstudio was founded in Florence in 1966–67, including the architects, Adolfo Natalini, Cristiano Toraldo di Francia, Roberto Magris, Piero Frassinelli, Alessandro Magris and Alessandro Poli; Lamas, A. Lamas, "50 Years of Architecture: Superstudio 50 Exhibition" MAXXI-Rome, in *Metalocus*, July 3, 2016, https://www.metalocus.es/en/news/50-years-architecture-superstudio-50, Retrieved: January 17, 2021.

[77] Ibid.

Fig. 11.10 Camp, Fundamental Acts, Superstudio 1971–72. ([URL]: https://www.bagtazocollection.com/blog/2016/1/20/theory-study-italian-conceptual-architecture-superstudio-utopia)

friction between history and present, nature and artifice, humans and inorganic matter generated a new, live, political space that could provoke thought".[78] Their visual narratives generally portray "dry, precise, lofty and absolute, archaic and everlasting" characteristics.[79] In accordance with Meadows' report, but approaching to the scientifically presented reality from the other hand, and through their interrogative images, Superstudio reflected the possible future of the world and the alternative life models. Besides, the apocalyptical atmosphere was also enhanced with characteristic tectonics reflecting from the images such as with evolving mega grid, or the continuous monument overarching the built and the unbuilt environments, or the super surface where it covers almost everything. In this way, these unconventional super planes proposed to facilitate an omnipresent interiority, on the ever-ending continuity of

[78] Ibid.

[79] Ibid.

interconnected planes that maintains every need of its citizen. The user unfolds or "super-fold" the grid to fulfil its needs just embedded to the surface. So from a Deleuzian perspective, this super surface almost became an *immanent plane of emergence*.

Parallel to Superstudio's provocative works, by the end of 1970s Rem Koolhaas' dystopias also emerged in the same way. Targeting the globalization this time and criticizing the local-global dialectic, Rem Koolhaas focused on the "urban density and the culture of congestion".[80] In this respect, in his book *Delirious New York: A Retroactive Manifesto* (1978), Rem Koolhaas proposed a "retrospective urban manifesto, which presents a critical and intellectual review of modern architecture and urbanism within the context of Manhattan".[81] Again by utilizing visuals (Fig. 11.11), Koolhaas raised his critics against the over-sized presence of the architecture, dominating and assimilating the minor spatialities.[82]

Briefly, the dystopic envisions of the twentieth century were critical stances to the occurring decay or the crises being as the consequences of greedy capitalism, cold-war militarization of the globe, emerging natural catastrophic events and drastic deficiencies of natural resources. Evidently, the projects mostly instrumentalized the interrogative impact of provocative images. These striking visual narratives criticized and projected the imperceptible aftermath of the impasses, mostly by reflecting the undesirable side of the capitalism. Despite reflecting the dark precursor of the movement, in fact the late twentieth-century dystopias challenged for the change and tried to increase the awareness of the society.

[80] R. Koolhaas, *Delirious New York: A Retroactive Manifesto*, Oxford University Press, 1978; "Delirious New York: A Retroactive Manifesto 1978", Designito: The Search Continuous. https://designito.net/2015/09/03/delirious-new-york-a-retroactive-manifesto-1978/ Retrieved: January 20, 2021.

[81] Ibid.

[82] Ibid.

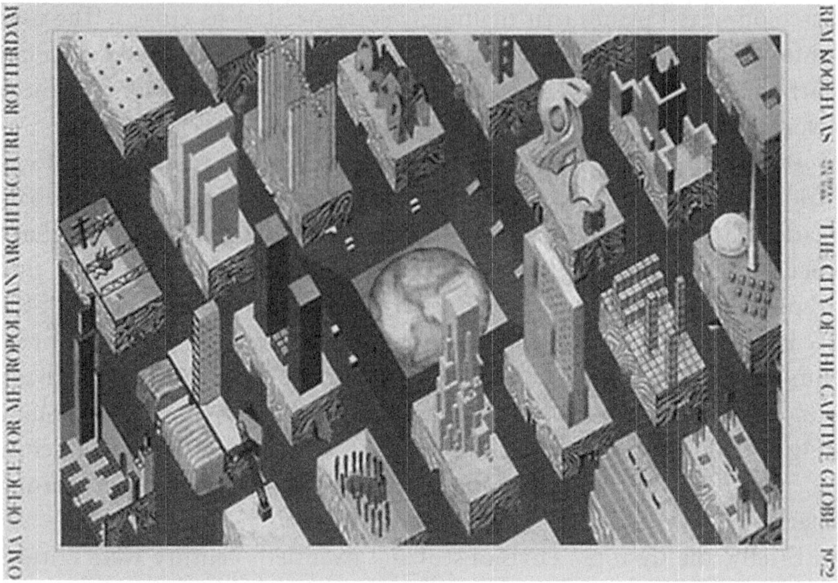

Fig. 11.11 City of the Captive Globe Project, Rem Koolhaas, 1972. (MIT Architecture, Urban Planning, and Visual Arts, Online Image Collection, https://dome.mit.edu/handle/1721.3/21258?show=fu)

Concluding Remarks: From Broader Inclusiveness Through Restrictive Exclusions, and the Next

Departing from the triggering potentials of the recent events of marginal violence and the assault on the contemporary urban realm, as well as the drained physical infrastructures of the city under the threat of the globally affecting diseases, this chapter revealed the continuous cycles of urban transformation, the heydays and the descent of the metropolitan condition significantly by focusing on the last hundred years of the modern era. Meanwhile, the intertwined relations of technology, politics and socio-economic factors and their transformative potentials in metropolitan life were revealed in relation to the human condition, and our configuration of public and private realms.

Since the progressive pattern of modernization recognized the public share as a democratic milieu of equally accessible realm for each and every existence,[83] a continuous plane of colliding vectors[84] shaping and being shaped with, a web of multi-discursive trajectories,[85] this aspect further consecrated the modern metropolis as an edificatory realm, the bedrock of mobility,[86] communication and cultural exchange with its related infrastructures. Despite the inclusive ideals of the modern and progressive metropolitan condition in the modernist agenda, individuals' interaction with growing metropolis mostly remained restricted due to the exclusive mechanisms of affordance in socio-economic terms, alienations in technological and social terms or oppressive in political and ideological terms. In this regard, the unpredictable, uneven presence of the public sphere encompassed the uncanny and multiple threats of discrimination, assault, violence even diseases, so that private realm became the shifting correspondence of the public sphere. On the other side, the two conflicting connotations of current private realm also appeared to be interrogative to be dismantled further in prospect studies. In this sense, while the affirmative connotation of interiority assigned a privilege for the insider being inclusive to the milieu that it existed and addressing a comfort zone; but it also proposed restrictive capacities for confining mechanism of either political stratification and homogenization, or even acting literally as detention rooms of the failing metropolis under the dominance of regressive ideologies.

Nevertheless, as reflected, continuous repeating cycles of the rise and fall of the urban condition, of the decline and the advancements, clearly keep flaming the light of hope, whereas the current decaying condition of the metropolitan milieu is evidently subject to change no matter how despairing it would be.

---

[83] O. Negt and A. Kluge, *Public Sphere and Experience: Toward an Analysis of the Bourgeois and Proletarian Public Sphere*, 1993; H. Arendt, *Human Condition*, 1958.

[84] G. Simmel, Metropolis and Mental Life, 1903.

[85] G. M. Furtado and C. Lopes, "Interpreting the Contemporary Metropolis: Notes on the Urban Debate and on Ignasi Solà-Morales", in *Footprint Journal*, Vol: 3 No: 5 Metropolitan Form, Autumn 2009, pp. 161–172.

[86] I. Solà-Morales, "Liquid Architecture", in *Anyhow*, Cynthia Davidson (ed.) The MIT Press, NY, 1998, pp. 36–43.

References

Adalet, B. 2018. *Hotels and Highways: The Construction of Modernization Theory in Cold War Turkey*. Stanford University Press.

Ahmed, M. 2015. Why Does ISIS Destroy Historic Sites? Centre on Religion & Geopolitics. *Tony: Blair Faith Foundation Web Archive*, September 1. https://web.archive.org/web/20151006100612/http://tonyblairfaithfoundation.org/religion-geopolitics/commentaries/opinion/why-does-isis-destroy-historic-sites. Accessed 7 November 2020.

Akpınar, I.Y. 2010. The Making of a Modern Pay-I That in Istanbul: Menderes Executions After Prost's Plan. In *The Imperial Capital to the Republican Modern City: Henri Prost's Planning of Istanbul (1936–1951)*, ed. C. Bilsel and P. Pinon, 167–199. Istanbul: İstanbul Research Institute.

Arendt, H. 1958. *Human Condition*. Chicago: University of Chicago Press.

Avermaete, T. 2014. Architecture as Res Publica: The Public Realm as Frame, Substance and Goal for Design. In *Delft Lecture Series Architectural Design*, ed. Susanne Komossa and Roberto Cavallo, 28–45. Delft: TU Delft Press.

Bland, L.I. Marshall and the Plan. The George C. Marshall Foundation. https://www.marshallfoundation.org/marshall/the-marshall-plan/history-marshall-plan/marshall-and-the-plan-by-larry-i-bland/. Accessed 18 January 2021.

Cellini, B. 2010. *The Autobiography of Benvenuto Cellini* (org. 1562). Penguin Books.

Childe, V.G. 1950. The Urban Revolution. In *Town Planning Review*, vol. 21, no. 1. Liverpool University Press. https://www.jstor.org/stable/i40003353. Accessed 10 November 2020; V.G. Childe. Man Makes Himself. London: Watts & Co., 1936.

Cook, P., ed. 1972. *Archigram*. London.

Crafts, N., and G. Toniolo. 2012. Les Trente Glorieuses: From the Marshall Plan to the Oil Crisis. Oxford Handbooks Online. https://doi.org/10.1093/oxfordhb/9780199560981.013.0018. Accessed 15 July 2020.

Currey, A. 2015. Here Are the Ancient Sites ISIS Has Damaged and Destroyed. *National Geographic*, September 1. Washington, DC: National Geographic Society. https://www.nationalgeographic.com/news/2015/07/150702-ISIS-Palmyra-destruction-salafism-sunni-shiite-sufi-Islamic-State/. Accessed 7 November 2020.

Curry, A. 2008. Gobekli Tepe: The World's First Temple? *Smithsonian Magazine*. https://www.smithsonianmag.com/history/gobekli-tepe-the-worlds-first-temple-83613665/. Accessed 7 November 2020.

Fields, N. 2006. *Ancient Greek Fortifications 500–300 BC (Fortress)*. Osprey Publishing; F. Yegül and D. Favro. *Roman Architecture and Urbanism: From the Origins to Late Antiquity*. Cambridge University Press, 2019.

Foursatié, J. 1979. *Les Trente Glorieuses, ou la Révolution Invisible de 1946 à 1975*. Paris: Fayard.

Fraser, N. 1992. Rethinking the Public Sphere: A Contribution to the Critique of Actually Existing Democracy. In *Habermas and the Public Sphere*, ed. Craig Calhoun. Cambridge, MA: MIT Press.

Furtado, G.M., and C. Lopes. 2009. Interpreting the Contemporary Metropolis: Notes on the Urban Debate and on Ignasi Solà-Morales. *Footprint Journal* 3 (5, Autumn) Metropolitan Form: 161–172.

Habermas, J. 1989. *The Structural Transformation of the Public Sphere: An Inquiry into a Category of Bourgeois Society*. Cambridge, MA: MIT Press.

Hahn, J. 2013. Labyrinth, the Shape of the Modern Mind: Kafka, Auster, Borges, Wellesley College, Honors. In *Comparative Literature*, 55–77. Wellesley College Digital Scholarship and Archive. http://repository.wellesley.edu/cgi/viewcontent.cgi?article=1143&context=thesiscollection. Accessed 1 July 2020.

Hobsbawm, E. 1995. *Age of Extreme: The Short Twentieth Century 1919–1991*. London: Abacus Book.

Howard, E. 1902. *Garden Cities of To-morrow*. London: S. Sonnenschein & Co. http://library.lol/main/59EB59F5FD9AC4B783C0C2EBE6BB7E76. Accessed 14 January 2021. D. Wiebenson. Tony Garnier: The cité industrielle (Planning and Cities). *Studio Vista* (January 1, 1969).

Kallipoliti, L. 2013. *Mission Galactic Household: The Resurgence of Cosmological Imagination in the Architecture of The 1960s and 1970s*. Princeton University, ProQuest Dissertations Publishing, 3560318. https://search.proquest.com/openview/30e913038f7988ab40e46dee85d9d560/1?pq-origsite=gscholar&cbl=18750&diss=y. Accessed 20 January 2021.

———. 2014. Cloud Colonies: Electronic Urbanism and Takes Zenetos' City of the Future in the 1960s. ACSA Annual Meeting Proceedings. https://www.acsa-arch.org/proceedings/Annual%20Meeting%20Proceedings/ACSA.AM.102/ACSA.AM.102.77.pdf. Accessed 16 January 2021.

———. 2020. AA Files No. 77, pp. 18–27.

Kodalak, G. 2015. A Monstrous Alliance: Open Architecture and Common Space. *Footprint Journal* 9 (1, Spring), Commoning as Differentiated Publicness: 69–89. https://doi.org/10.7480/footprint.9.1.900. Accessed 16 January 2021.

Koolhaas, R. 1978. *Delirious New York: A Retroactive Manifesto*. Oxford University Press.

Kwinter, S. 2002. *Architectures of Time: Toward a Theory of the Event in Modernist Culture*. Cambridge, MA: MIT Press.

Lamas, A. 2016. 50 Years of Architecture: Superstudio 50 Exhibition. MAXXI-Rome. *Metalocus*, July 3. https://www.metalocus.es/en/news/50-years-architecture-superstudio-50. Accessed 17 January 2021.

Langer, W. 1972. *An Encyclopedia of World History*. 5th ed. Boston, MA: Houghton Mifflin Company.

Mathews, S. 2005. The Fun Palace: Cedric Price's Experiment in Architecture and Technology. *Technoetic Arts: A Journal of Speculative Research* 3 (2). Intellect Ltd. https://doi.org/10.1386/tear.3.2.73/1.

Mattsson, H. 2015. Were the Motorways Meet: Architecture and Corporatism in Sweden 1968. In *Architecture and the Welfare State*, ed. M. Swenarton, T. Avermaete, and D. van der Heuvel, 155–175. New York: Routledge.

Meyer, D.R. 1990. The Rise of the Industrial Metropolis: The Myth and the Reality. *Social Forces* 68 (3): 731–752. https://www.jstor.org/stable/2579351. Accessed 30 October 2020.

Negt, O., and A. Kluge. 1993. *Public Sphere and Experience: Toward an Analysis of the Bourgeois and Proletarian Public Sphere*. Minneapolis: University of Minnesota Press.

Oliver, D., and J. Notroff. 2015. A Sanctuary, or So Fair a House? In Defense of an Archaeology of Cult at Pre-Pottery Neolithic Gobekli Tepe. In *Defining the Sacred: Approaches to the Archaeology of Religion in the Near East*, ed. N. Laneri. Oxbow Books.

Pask, G. The Architectural Relevance of Cybernetics. In *Computational Design Thinking*, 68–77. https://cmusyntheticecologies.files.wordpress.com/2014/01/pask_gordon_thearchitectural-relevance-of-cybernetics.pdf. Accessed 10 January 2021.

"Peter Cook (Archigram): Instant City, 1968–1970". FRAC Centre-Val de Loire. https://www.frac-centre.fr/_en/art-and-architecture-collection/cook-peter/instant-city-317.html?authID=44&ensembleID=113. Accessed 13 December 2020.

Pimlott, M. 2014. Fiction and Significance in the Public Interior: Culture & (Its) Imagination: Spaces and Methods of Place Making. In *Delft Lecture Series Architectural Design*, ed. Susanne Komossa and Roberto Cavallo, 28–45. Delft: TU Delft Press.

Redekop, E. 1980. Labyrinths in Time and Space. *Mosaic: An Interdisciplinary Critical Journal* 13 (3/4). Other Worlds: Fantasy and Science Fiction Since

1939 (Spring/Summer). University of Manitoba, pp. 95–113, https://www.jstor.org/stable/24780265. Accessed 16 January 2021.
de Regil, A.J. 2001. Keynesian Economics and the Welfare State. *Neo-Capitalist Assault*, the Jus Semper Alliance, N:4, pp. 1–19. https://www.jussemper.org/Resources/Economic%20Data/The%20Neo-Capitalist%20Assault/Resources/KeynesianEconomics.pdf. Accessed 16 January 2021.
River, C. 2019. *The Incense Trade Route: The History of the Rise and Fall of an Ancient Global Economy*. Charles River Editor; R. McLaughlin. *The Roman Empire and the Silk Routes: The Ancient World Economy and the Empires of Parthia, Central Asia and Han China*. Pen and Sword History, 2020.
Romey, K. 2015. ISIS Destruction of Ancient Sites Hits Mostly Muslim Targets. *National Geographic*, July 2. Washington, DC: National Geographic Society. https://www.nationalgeographic.com/news/2015/07/150702-ISIS-Palmyra-destruction-salafism-sunni-shiite-sufi-Islamic-State/. Accessed 7 November 2020.
Simmel, G. 1903. Metropolis and Mental Life. In *Simmel on Culture: Selected Writings*, ed. D.P. Frisby and M. Featherstone. SAGE Publications Ltd.
Smith, M.E. 2009. V. Gordon Childe and the Urban Revolution: A Historical Perspective on a Revolution in Urban Studies. *Town Planning Review* 80 (1). Liverpool University Press. https://patterns.architexturez.net/system/files/MES-09-Childe-TPR.pdf. Accessed 10 November 2020.
Smithson, A., and P. Smithson. 1958. Mobility: Road System. *Architectural Design*, October.
———, ed. 1968. *Team 10*. London: Studio Vista; M. Risselda and D. van den Heuvel, eds. *Team 10*. Rotterdam: NAİ Publishers, 2005.
Solà-Morales, I. 1998. Liquid Architecture. In *Anyhow*, ed. Cynthia Davidson, 36–43. New York: The MIT Press.
Tafuri, M. 1979. *Architecture and Utopia: Design and Capitalist Development*. Boston, MA: MIT Press.
Torisson, F. 2017a. The Cybernetic Hypothesis & Architecture. *Histories of PostWar Architecture* (1). https://doi.org/10.6092/issn.2611-0075/7208. Accessed 18 January 2021.
———. 2017b. *Utopology a Re-interrogation of the Utopian in Architecture*. Doctoral Dissertation in Architecture, Lund University.
UN Department of Economic and Social Affairs Population Dynamics. 2017 Revision of UN World Urbanization Prospects. https://population.un.org/wup/. Accessed 30 October 2020.
UN World Heritage Convention. Historic Centre of San Gimignano. https://whc.unesco.org/en/list/550. Accessed 30 October 2020.

Vidler, A. 1993. Spatial Violence. *Assemblage*, No. 20, pp. 84–85.
Wallerstein, I. 2011. *The Modern World-System II: Mercantilism and the Consolidation of the European World-Economy, 1600–1750*. University of California Press.
Whiteley, N. 1987. Toward a Throw-Away Culture. Consumerism, 'Style Obsolescence' and Cultural Theory in the 1950s and 1960s. *Oxford Art Journal*. https://ccnymadmen.files.wordpress.com/2016/07/toward-athrow-away-culture.pdf. Accessed 18 January 2021.
Wiebenson, D. 1960. Utopian Aspects of Tony Garnier's Cité Industrielle. *Journal of the Society of Architectural Historians* 19 (1, Mar.). University of California Press. https://www.jstor.org/stable/987962. Accessed 14 January 2021.

Index[1]

[1] Note: Page numbers followed by 'n' refer to notes.

S. S. Das, A. R. Pratihar (eds.), *Technology, Urban Space and the Networked Community*, https://doi.org/10.1007/978-3-030-88809-1

The manufacturer's authorised representative in the EU is Springer Nature Customer Service Centre GmbH, Europaplatz 3, 69115 Heidelberg, Germany. If you have any concerns regarding our products, please contact ProductSafety@springernature.com

Printed and bound by CPI Group (UK) Ltd, Croydon, CR0 4YY
30/09/2026
02225490-0009